Characterizing the Uncrewed Systems Industrial Base

BRADLEY WILSON, ELLEN M. PINT, ELIZABETH HASTINGS ROER,
EMILY ELLINGER, FABIAN VILLALOBOS, MARK STALCZYNSKI,
JONATHAN L. BROSMER, ANNIE BROTHERS, ELLIOTT GRANT

Prepared for the OUSD(R&E)
Approved for public release; distribution unlimited

RAND NATIONAL DEFENSE RESEARCH INSTITUTE

For more information on this publication, visit **www.rand.org/t/RRA1474-1**.

About RAND

The RAND Corporation is a research organization that develops solutions to public policy challenges to help make communities throughout the world safer and more secure, healthier and more prosperous. RAND is nonprofit, nonpartisan, and committed to the public interest. To learn more about RAND, visit www.rand.org.

Research Integrity

Our mission to help improve policy and decisionmaking through research and analysis is enabled through our core values of quality and objectivity and our unwavering commitment to the highest level of integrity and ethical behavior. To help ensure our research and analysis are rigorous, objective, and nonpartisan, we subject our research publications to a robust and exacting quality-assurance process; avoid both the appearance and reality of financial and other conflicts of interest through staff training, project screening, and a policy of mandatory disclosure; and pursue transparency in our research engagements through our commitment to the open publication of our research findings and recommendations, disclosure of the source of funding of published research, and policies to ensure intellectual independence. For more information, visit www.rand.org/about/research-integrity.

RAND's publications do not necessarily reflect the opinions of its research clients and sponsors.

Published by the RAND Corporation, Santa Monica, Calif.
© 2023 RAND Corporation
RAND® is a registered trademark.

Library of Congress Cataloging-in-Publication Data is available for this publication.
ISBN: 978-1-9774-0975-1

Cover: Air: U.S. Marines with Marine Unmanned Aerial Vehicle Squadron 2 launch a RQ-21A Blackjack during Frozen Reindeer in Rena, Norway, May 21, 2018. Frozen Reindeer is a training exercise designed to build flight and ground proficiency to increase unit readiness. (U.S. Marine Corps photo by Sgt. Joselyn Jimenez).

Surface: The unmanned surface vessel Powervent, with the U.S. Navy's Naval Surface Warfare Center Combatant Craft Division, speeds across the water while equipped with cameras, computer systems and non-lethal weapons during the Spiral 1 experiment for Trident Warrior 2012 at Fort Eustis, Va., Jan. 31, 2012. Trident Warrior, sponsored by Commander U.S. Fleet Forces Command, is a program that experiments advanced maritime initiatives in an operational environment to improve capabilities available to the fleet. (U.S. Navy photo by Mass Communication Specialist 3rd Class Betsy Knapper).

Ground: A U.S. Army Pacific soldier walks down a trail 22 July 2016 while controlling an unmanned vehicle as part of the Pacific Manned Unmanned–Initiative at Marine Corps Training Area Bellows, Hawaii. (Photo by Staff Sgt. Christopher Hubenthal, U.S. Air Force).

Subsurface: Members of the Unmanned Underwater Vehicle (UUV) detachment, Commander, Task Group (CTG) 56.1, guide a UUV as it is lowered into the water during a training exercise. U.S. Navy photo by Mass Communication Specialist 1st Class Peter Lewis.

About This Report

The Office of the Under Secretary of Defense for Research and Engineering (OUSD[R&E]) asked the RAND Corporation to review the posture of the defense industrial base (DIB) to produce and sustain required levels of autonomous uncrewed systems (UxS) and asked for recommendations for strengthening the U.S. autonomous UxS industrial base and supply chain and for increasing capacity for the production and sustainment of UxS inventory. We therefore provide an analysis of required levels of UxS to meet DIB requirements, identify and collect data on DIB categories and specific component elements within those categories, assess the posture of the DIB to produce and sustain levels of UxS platforms as required by the U.S. Department of Defense (DoD), conduct a comparative analysis of near-peer nation-states China and Russia, and provide recommendations on strengthening the DIB.

The research reported here was completed in March 2022 and underwent a security review with the sponsor and the Defense Office of Prepublication and Security Review before public release.

The RAND National Security Research Division

This research was sponsored by OUSD(R&E) and conducted within the Acquisition and Technology Policy Center of the RAND National Security Research Division, which operates the National Defense Research Institute, a federally funded research and development center sponsored by the Office of the Secretary of Defense, the Joint Staff, the Unified Combatant Commands, the U.S. Navy, the U.S. Marine Corps, the defense agencies, and the defense intelligence enterprise.

For more information on the RAND Acquisition and Technology Policy Center, see www.rand.org/nsrd/atp or contact the director (contact information is provided on the webpage).

Acknowledgments

We thank our sponsor Bethany Harrington, director, Technology and Industrial Base Assessments in OUSD(R&E), and her colleague Rachel Major for continuous feedback during the project. At the Office of the Director of Defense Research and Engineering for Modernization, we thank Wayne Nickols, principal director for autonomy, and his successor, Jaret C. Riddick, for acting as catalysts in guiding the scope of this research. We are also grateful to Mike Di Paolo for his robust feedback throughout the project.

Although we cannot thank them by name, there are numerous other DoD personnel and industry partners with whom we met who were instrumental in sharing data and information.

We thank our RAND colleagues Michael Kennedy and Scott Savitz for their technical reviews, which helped improve the quality of this report.

Summary

Motivation

The Office of the Under Secretary of Defense for Research and Engineering (OUSD[R&E]) has hypothesized that the demand signal for uncrewed systems (UxS) in the coming years will strain the capacity of the U.S. defense industrial base (DIB). OUSD(R&E) asked the RAND Corporation to explore this possibility and assemble relevant risks, issues, and opportunities to support the office's ongoing activities, including complying with statutory requirements to provide annual reports to the U.S. Congress under Section 217 of the National Defense Authorization Act for fiscal year (FY) 2020.

The scope of the request covered many types of UxS, which include uncrewed aerial systems (UAS), uncrewed ground systems (UGSs), and maritime platforms—specifically, uncrewed surface vehicles (USVs) and uncrewed underwater vehicles (UUVs). Additionally, OUSD(R&E) asked for a comparison of the U.S. UxS DIB to relevant near-peer nations of Russia and China, and other global players as warranted.

Approach

We provide an analysis of required levels of autonomous UxS to meet DIB requirements, identify and collect data on DIB categories and specific component elements within those categories, assess the posture of the DIB to produce and sustain levels of UxS platforms as required by the U.S. Department of Defense (DoD), conduct a comparative analysis of near-peer nation-states China and Russia, and provide recommendations on strengthening the DIB.

To accomplish this, we consulted a variety of data sources, such as the Association for Uncrewed Vehicle Systems International (AUVSI) market reports and robotics database, DoD budget documents, and the Janes Markets Forecast to characterize the UxS demand signal in terms of expected procurement units and costs. To supplement our analyses, we conducted interviews with officials from U.S. military UxS program offices, representatives from commercial entities, and subject matter experts (SMEs).

To help us characterize and scope UxS platform suppliers and components, we developed sample sets of 18 UxS platforms produced by companies in the United States, China, and Russia, and one containing a mix of platforms produced by other countries. We also developed a Work Breakdown Structure to identify important categories of UxS subsystems. We used these categories in conjunction with the sample sets to guide an exploration of the component technologies and associated firms. Data on component subsystems are not available in the aggregate and are otherwise time-consuming to assemble.

To characterize the capacity of the DIB, we conducted fragility and criticality assessments. *Fragility* refers to the likelihood of a capability being lost; *criticality* refers to the extent to which a capability (e.g., technology, part, component, product) is difficult to replace if lost. In conducting these assessments, we looked at fragility indicators, such as financial outlook, DoD sales, the number of firms in relevant sectors, and dependence on foreign sources of supply. We also looked at criticality indicators such as defense uniqueness, design requirements, skilled labor needs, facility and equipment availability, materials and components with long lead times, and availability of alternatives.

Findings

Overall we found that expected U.S. military demand was increasing from FY 2021 to FY 2026 for all types of UxS (except for UUVs, the demand for which declines slightly). Although the scale of the expected increase varied significantly by the type of system, the increase is unlikely to burden the DIB by FY 2026. Only UAS unit demand was expected to increase more than 100 percent over FY 2021 levels—driven primarily by higher demand for small, low-cost platforms between FY 2024 and FY 2026. Military UGS unit demand increased substantially from FY 2019 to FY 2021 and was expected to increase 58 percent through FY 2026, with demand peaking in FY 2024. However, total cost growth over the same period (FY 2021–FY 2026) was expected to be much larger, at 460 percent. Military USV and UUV unit demand was expected to remain essentially flat from FY 2021 to FY 2026, though this would still represent a modest increase over demand levels in FY 2018 and earlier.

Overall, estimated U.S. military procurement of UxS in FY 2021—excluding research, development, test, and evaluation—ranged from $1.6 billion to $3.3 billion. FY 2026 estimates ranged from $4.8 billion to $8.4 billion, though there is likely significant uncertainty in these estimates.

Compared with albeit limited data on commercial sales volumes, the U.S. military UAS demand signal is low: demand in FY 2022 ranged from 1,200 to approximately 3,000 units, suggesting that U.S. military demand for UAS equated to roughly 1–3 percent of overall yearly U.S. commercial demand. UAS market data are highly variable, and while DoD seems like a small player on a per-unit level, it is a larger player on a per-dollar basis (13–70 percent of the North American UAS market).

We find that the U.S. UxS DIB is more fragile than it is critical—that is, there is more risk to losing capabilities than there is in replacing capabilities once they are gone. This is due to financial outlook risks; maritime demand uncertainty; a limited number of firms building large UGSs, USVs, and UUVs; and foreign dependence on selected components (e.g., lithium-ion batteries). There are also criticality issues in a lack of defense-ready commercial off-the-shelf (COTS) UGSs, USVs, and UUVs and the availability of alternative domestic sources of materials and components such as lithium, carbon fiber, titanium, and semiconductors. However, none of these risks seems sufficient to be very concerned about the capacity of the DIB to produce and sustain expected U.S. military UxS demand levels. These risks and issues would be exacerbated in the event expected demand levels increase significantly, such as a policy shift to wartime production needs.

Data limitations make it challenging to compare Chinese, Russian, and U.S. demand and capacity. We are relying on open sources, and the information may be unreliable. With these caveats, we have identified some general trends. First, the military demand signals in Russia and China both significantly lag expected U.S. military demand in terms of UAS units by roughly a factor of five, as shown in Figure S.1.

Figure S.1. U.S., Chinese, and Russian Historical and Expected Military UAS Units Delivered, 2016–2026

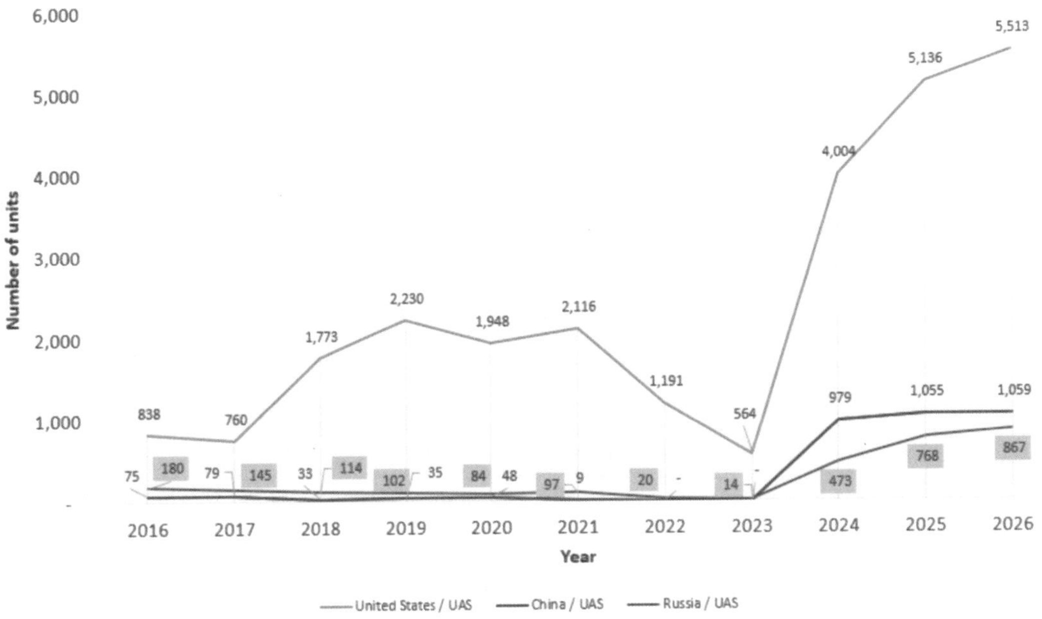

SOURCE: Authors' analysis of data from Janes Markets Forecast in May 2022.
NOTE: Russian unit data labels have a blue background.

Both China and Russia are expected to have significantly lower demand for UGSs than the United States, even relative to historical trends. Expected Chinese and Russian military demand for both USVs and UUVs also lags behind U.S. demand, as shown in Figure S.2.

UAS appear to be the only system types with significant current Chinese capabilities, whereas UGSs, USVs, and UUVs all require further development before being employed more commonly. Chinese sources have cited safety, accuracy, and interoperability issues, among others, with these non-UAS. A positive for China is that it has the least defense-unique UAS industrial base of the three primary countries we examined, as Figure S.3 shows.

Figure S.2. U.S., Chinese, and Russian Historical and Expected Military USV and UUV Units Delivered, 2016–2026

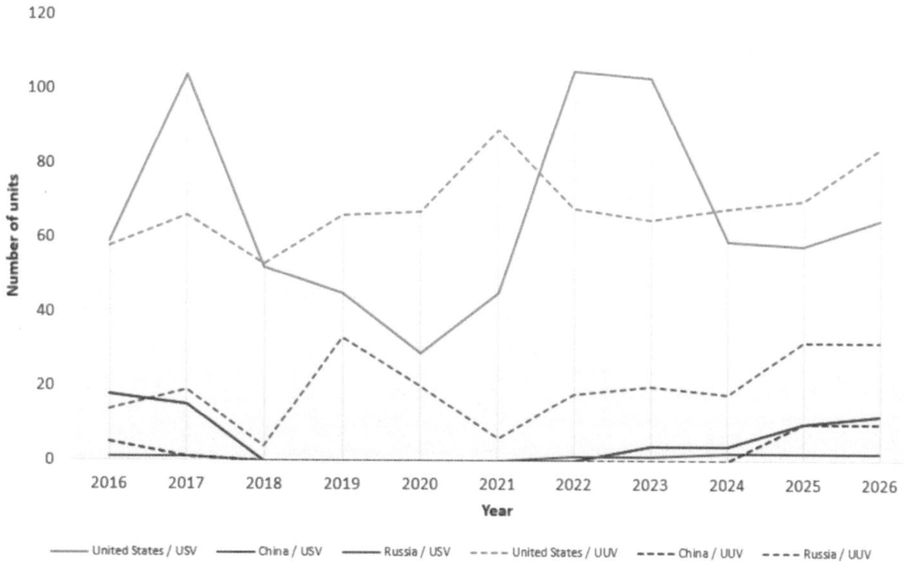

SOURCE: Authors' analysis of data from Janes Markets Forecast.
NOTES: The scale has changed from Figure S.1. UUVs are shown with a dashed line.

The percentage of UAS platforms used for defense are China, 27 percent; United States, 54 percent; and Russia, 86 percent. The United States has the least defense-unique UGS, USV, and UUV markets (by this measure).

In general, Russia has clearly demonstrated advancements in its uncrewed technologies. Although it might not be a major exporter of uncrewed platforms now, it remains the second-largest exporter of arms after the United States. At the time of this writing, Russia had a few contracts with Myanmar and unspecified countries for UAS exports, and it was not clear how Russia's invasion of Ukraine would affect these arrangements or future contracts. Notably, the deal with Myanmar was for the Orlan-10E (export version); the Orlan-10 was found to contain many foreign components. In the short term, the largest procurer of Russian uncrewed platforms will remain the Russian Ministry of Defense.

Russian Ground Forces have employed UAS in combat in Ukraine,[1] including the Granat (produced by the Kalashnikov Concern, a subsidiary of the state-owned enterprise Rostec) and the Orlan-10 (produced by the Special Technology Center, a Russian contractor).

---

[1] Sten Allik, Sean Fahey, Tomas Jermalavičius, Roger McDermott, and Konrad Muzyka, *The Rise of Russia's Military Robots: Theory, Practice and Implications*, Tallinn, Estonia: International Centre for Defence and Security, February 2021.

Figure S.3. Comparison of Defense Uniqueness

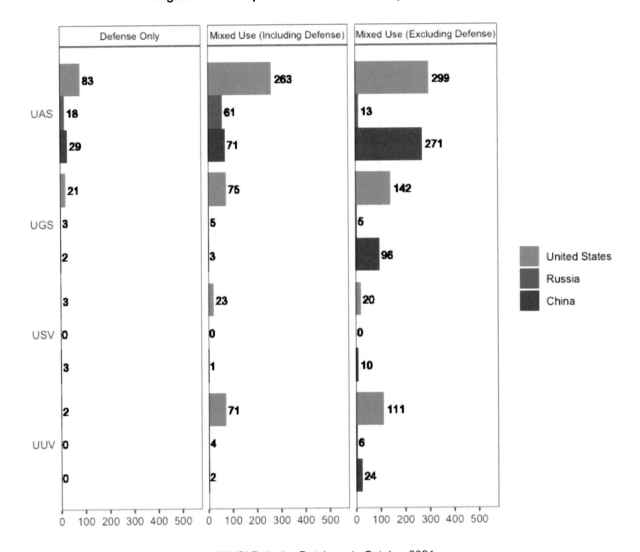

SOURCE: Authors' analysis of data from AUVSI Robotics Database in October 2021.

Further, the Russian UxS industrial base seems to lack key assemblies, such as sophisticated electronics, and it relies heavily on China for engine components. Machine shops and plants might still depend on foreign-built machine tools. U.S. sanctions following Russia's invasion of Ukraine in 2022 restricted key technological exports to Russia, such as semiconductors, aviation parts, lasers, and sensors produced in foreign countries (e.g., Taiwan) that rely on software and equipment developed in the United States. The United States exported only $114 million in microchips to Russia in 2021, and Russia may turn to China for substitutes. Even so, the lack of access to these products will negatively affect the Russian uncrewed industrial base over time.

Recommendations

Our analyses and interviews with stakeholders and SMEs led to the following recommendations for OUSD(R&E) as it considers its options to mitigate risk to the UxS DIB going forward:

1. Improve demand certainty for USVs and UUVs to mitigate the fragility of the maritime DIB.
2. Explore resilience mitigations for lithium-ion batteries, UAS motors, electro-optical components, semiconductors, light detection and ranging sensors, ferrite, iridium, titanium, and fiberglass. Caution is warranted because vendors indicated two common themes for global sourcing: low prices and higher quality (for selected components such as inertial navigation systems). Moving too quickly with resilience measures may have unintended consequences.
3. Expand the Defense Innovation Unit's Blue UAS project (or a similar effort) to include UGSs, USVs, and UUVs. This will create more defense-ready COTS systems and better enable smaller performers to engage DoD.
4. Explore options to boost the large UAS commercial industrial base to mitigate defense uniqueness in large UAS platforms. (Hurdles may exist due to Federal Aviation Administration policies.)
5. Support policies that grow important trade skills, such as electronics soldering (for UAS), welding, forging, and metal casting (for USVs and UUVs).

Contents

Figures and Tables

Figures

Tables

Chapter 1. Motivation and Background

The Office of the Under Secretary of Defense for Research and Engineering (OUSD[R&E]) asked the RAND Corporation to review the posture of the U.S. defense industrial base (DIB) to produce and sustain required numbers of autonomous uncrewed systems (UxS) and asked for recommendations for strengthening the U.S. DIB and supply chain for autonomous UxS and for increasing U.S. capacity to produce and sustain the required autonomous UxS inventory. For reasons we discuss in this chapter, our analysis and conclusions focus largely on UxS and do not distinguish autonomous from nonautonomous systems.

Motivation

OUSD(R&E) proposed a hypothesis that the potential demand signal for all UxS (autonomous and nonautonomous) in the coming years will strain the capacity of the DIB. We explored this hypothesis and assembled relevant risks, issues, and opportunities to inform the office's ongoing activities, such as providing annual reports to the U.S. Congress as required by Section 217 of the National Defense Authorization Act (NDAA) for fiscal year (FY) 2020. A guide on risks, issues, and opportunities management produced by the Office of the Deputy Assistant Secretary of Defense for Systems Engineering defines *risks* as "potential future events or conditions that may have a negative effect on achieving program objective," *issues* as "events or conditions with negative effect that have occurred," and *opportunities* as "potential future benefits to the program."[1]

The scope of the request covered many types of UxS, which include uncrewed aerial systems (UAS) and uncrewed ground systems (UGSs), as well as two maritime platforms: uncrewed surface vehicles (USVs) and uncrewed underwater vehicles (UUVs). Additionally, OUSD(R&E) asked for a comparison of the U.S. UxS DIB to relevant near-peer nations of Russia and China, and other global players as warranted.

Background

UxS are not a new concept or capability for DoD. However, like many capabilities, the technology is continually evolving and maturing, presenting potential new applications while also lowering cost. Additionally, UxS capabilities continue to find new, nondefense applications that have grown and diversified the industry outside DoD.

---

[1] Office of the Deputy Assistant Secretary of Defense for Systems Engineering, *Department of Defense Risk, Issue, and Opportunity Management Guide for Defense Acquisition Programs*, Washington, D.C., January 2017.

The maturation of sensors, batteries, computing power, software, and other technologies has ushered in systems that are increasingly capable of automation and even autonomous operation. DoD continues to advance a research agenda aimed at evolving these technologies while also seeing demand from Congress, military leadership, and operational end users to employ the systems in every domain.

The growth in the industry and advancements in technology have also put increasingly capable systems into the hands of peer and nonpeer competitors alike. This has led to a growing counter-UxS industry in the civil and military sectors as state and nonstate actors mature or access capable counter-UxS systems. The U.S. government and other UxS users are presented with a quandary: whether to continue to research and develop UxS that can evade counter-UxS systems or instead to develop lower-cost (and presumably less capable) systems. The lower-cost UxS tend to be smaller, more plentiful, and suitable for networked operation (e.g., as drone swarms) but have shorter life spans or are intentionally disposable, and The Chief of Naval Research has stated that a critically important component of future naval success

> is incorporating advanced cyber-physical technologies found in the "small, the agile, and the many"—small unmanned, autonomous platforms that have the agility to be built and adapted quickly, in large numbers, and at far lower costs compared to larger platforms. These unmanned air, surface and subsurface vehicles will carry an array of sensors and modern payloads, and perform multiple missions.[2]

Yet another concern arising from the increasing availability of these lower-cost UxS and components is that such systems are no longer solely the domain of well-funded militaries; they are now increasingly commercialized, commoditized, and available for purchase through vendors that support the private sector as much or more than the military market. There is concern in OUSD(R&E) that it will present problems for DoD if demand for UxS increases significantly, a scenario that would require sourcing from suppliers and supply chains outside DoD's sphere of influence. A robust U.S. commercial industrial base could provide additional sources of supply to DoD, but existing designs may be more reliant on foreign components than DoD would allow.

Secretary of Defense Lloyd Austin recently noted in a speech that he "saw unmanned, solar-powered, Navy vessels that use AI [artificial intelligence] to build a shared picture of the surrounding seas," but in the context of it taking "too long to get that kind of innovation to our warfighters."[3] As potential remedies he mentioned the option for the Defense Advanced Research Projects Agency (DARPA) to team with corporate leaders and investors, as well as increasing funding for the Small Business Innovation Research Program and opening Defense Innovation Unit (DIU) tech hubs in Chicago and Seattle. Austin also suggested eliminating

---

[2] Warren Duffie, Jr., "ONR Chief Unveils New Vision to Reimagine Naval Power," press release, Arlington Va.: Office of Naval Research, November 23, 2021.

[3] Lloyd J. Austin III, "Remarks by Secretary of Defense Lloyd J. Austin III at the Reagan National Defense Forum (as Delivered)," transcript, U.S. Department of Defense, December 4, 2021.

irrelevant programs and platforms and a rapid defense experimentation reserve fund as ways of making funds available to test new technologies and to potentially streamline acquisition processes.

Tasks and Scope

This study had six tasks that guided our analytic approach:

- Task 1: Identify required levels of UxS to meet DIB requirements.
- Task 2: Identify and collect data on DIB categories and specific elements within those categories.
- Task 3: Assess the posture of the DIB to produce and sustain DoD-required levels of UxS platforms.
- Task 4: Conduct a comparative analysis of near-peer nation-states China and Russia.
- Task 5: Document and evaluate risks, issues, and opportunities for the future state of the domestic UxS DIB using conclusions derived from tasks 3, 4, and 5.
- Task 6: Provide recommendations for mitigation strategies and opportunities to improve autonomous platform development and sustainment capacity.

The UxS DIB is expansive and challenging to characterize, and OUSD(R&E) was interested in an exploratory study that looked broadly across the following DIB elements: autonomy-enabling UxS technologies, fundamental UxS technologies, manufacturing and production, raw materials, the production workforce, and testing and evaluation. The following are some examples of these elements:

- **Autonomy-enabling UxS technologies** include specialized integrated circuits with application-specific integrated circuits, graphics processing units, and field-programmable gate arrays; navigation that does not employ the Global Navigation Satellite System (GNSS); secure/resilient command, control, communications, and computers; autonomous tool kits; and sensors.
- **Fundamental UxS technologies** include batteries, propulsion, and other technologies that have dependencies on nondomestic material, design, manufacturing, or workforces.
- **Manufacturing and production capabilities** include advanced composite manufacturing and additive manufacturing.
- **Raw materials** include alloys and rare earth elements.
- **The production workforce** includes both skilled and unskilled workers, as well as the availability of training and certifications.
- **Testing and evaluation** includes the availability of test ranges, among many other capabilities.

This list is not intended to be exhaustive. There is considerable overlap between autonomous and nonautonomous UxS elements. Our original scope included autonomous UxS and shared elements (as demonstrated in Figure 1.1) and excluded nonautonomous elements.

Figure 1.1. Autonomous, Nonautonomous, and Shared UxS Elements

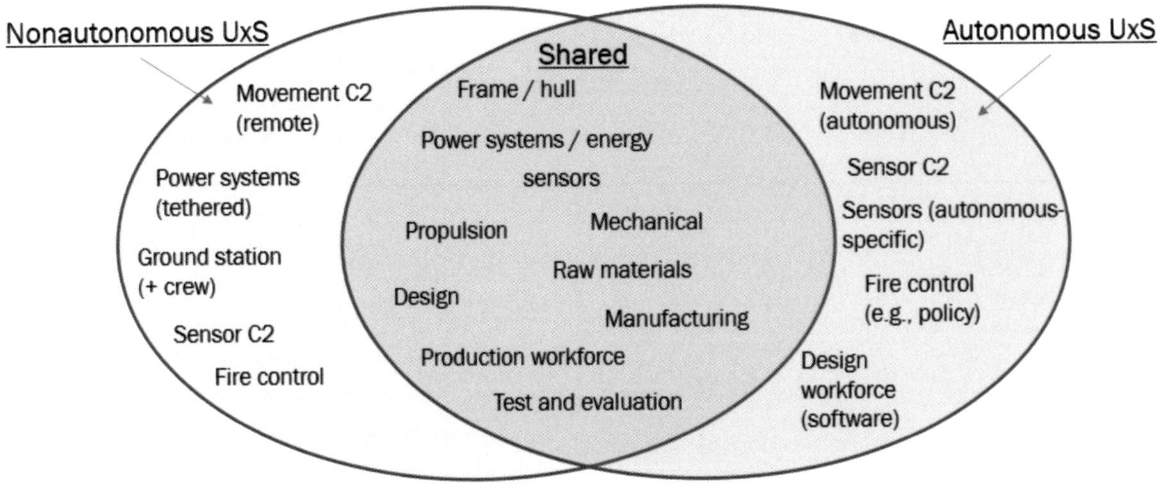

NOTES: The elements in this figure are a simplification and should not be considered exhaustive, because autonomy is generally viewed on a spectrum; C2 = command and control.

However, in practice, there are not separate industrial bases for autonomous and nonautonomous UxS. Further, the data sets we identified did not meaningfully distinguish between autonomous and nonautonomous systems. Therefore, our capacity analysis applies more broadly to the industrial base for UxS, though we did explore and inquire about autonomy-enabling technologies and components where possible.

Finally, it is also important to note that the timeline for this research was the near term—no longer than the five-year Future Years Defense Program.

Approach

This study had a large scope and was therefore exploratory in nature. We analyzed a variety of existing government and commercial data sets (e.g., from the Association for Uncrewed Vehicle Systems International [AUVSI], CrunchBase, FactSet, the Janes Markets Forecast, and government budgets) and conducted interviews with government and industry representatives to identify potential risks, issues, and opportunities in the UxS DIB.

Demand Signal

We used several approaches to characterize the DoD demand signal for UxS using DoD budget data, AUVSI budget data, and Janes Markets Forecast data, each described in the following sections.

DoD Budget Data

The U.S. government does not readily identify and organize UxS-related budget plans for research, development, test, and evaluation (RDT&E); procurement; and sustainment into a useful package for analysis. Therefore, to develop an understanding of the future short-term demand signal for the U.S. UxS DIB, we used a mixed methods approach to build a thorough scan of DoD budget lines across the military services for 2019 and 2020. We examined major DoD budget publications, service procurement documents, and third-party DoD budget analyses. We developed a data set listing each service's budget, UxS size category,[4] platform name, number of units ordered, expected year of delivery, and the specific budget document listing the quantity.

We built a series of known platform acquisitions, both historical and for the potential near-term future, for UxS across the U.S. Air Force, U.S. Army, and U.S. Navy. All of the Air Force's and most of the Army's procurements with budget lines containing deliverable units were for the air domain. The Navy had several non-UAS platforms, and the Army's UGS platforms included the Robotic Combat Support System, explosive ordnance disposal (EOD) robotics systems recapitalization, and other small robotic systems, but the Robotic Combat Vehicle was not mentioned in the Army's procurement budget documents.[5]

This method allowed us to confirm UxS procurement for the services, but it did not provide a full picture of the UxS demand signal. Specifically, we encountered budget lines that were missing information, a lack of a clear definitions of UxS, and abbreviated terminology in the budget documents. Because of the project's time constraints, we could not address many of these issues; therefore, we used additional data sources to supplement our analyses of UxS budgets.

Association for Uncrewed Vehicle Systems International Budget Data

AUVSI is an international nonprofit organization that gathers information on UxS.[6] The organization releases an analysis of DoD budget documents annually and provides data on the funding each service has allocated for UxS procurement and RDT&E. The U.S. government's FY 2021 budget and other sources mention funding for UxS for military purposes, but there is not currently aggregate funding for uncrewed projects across the military services.[7] AUVSI compiles data annually on all UxS across the U.S. military—more than 1,000 individual projects at different stages of development and with different funding requirements.[8] For FY 2021,

---

[4] These size categories are defined later in chapter when we describe the sample sets that we developed to support our analyses.

[5] The Army's RDT&E budget justification books for FY 2022 mention several ground systems that are in development, but do not provide information about the number of units being produced.

[6] AUVSI, "Who Is AUVSI?" webpage, undated b.

[7] Ronald O'Rourke, *Navy Large Unmanned Surface and Undersea Vehicles: Background and Issues for Congress*, Washington, D.C.: Congressional Research Service, R45757, October 20, 2021; Office of Management and Budget, *A Budget for America's Future: Budget of the U.S. Government, Fiscal Year 2021*, Washington, D.C.: U.S. Government Publishing Office, 2020.

[8] AUVSI, "Funding for Unmanned Systems in the FY 2022 Defense Budget Request," *Assuring Autonomy Blog*, June 16, 2021b.

AUVSI reported that the defense UxS RDT&E and procurement budget was $7.5 billion, roughly 1 percent of the DoD budget.[9] In searching various sources to find alternative numbers to confirm this data, it was difficult to find another comprehensive breakdown of UxS budgets. Given that AUVSI did not publish its own calculations, it is likely that some of these figures are linked to confidential sources (such as certain projects).

Janes Markets Forecast Data

The final source we used for the demand signal analysis was Janes Markets Forecast for UxS. Janes describes itself as an open-source defense intelligence company that provides data on military capabilities by cross-referencing government documents, company reports, news articles, forecasts, and public records.[10] The Janes data were the most comprehensive that we obtained for the project; however, they do not represent a complete list of UxS. For example, some data lines referred only to stated or derived opportunities, with limited populated fields—a result of incomplete source data. We performed several analyses to ensure that we interpreted and represented the Janes data correctly in this report, including validating sources, cross-checking budget lines, verifying consistency, and communicating with Janes representatives to resolve questions. All data used in the tables and figures that cite Janes Markets Forecast were collected directly from Janes and organized and interpreted to capture the U.S. UxS demand signal in the near term.

In gathering the Janes data, we imposed the following exclusions and assumptions:

- All the data were restricted to uncrewed platforms, including those that are optionally manned.
- The user type for all units was dual-use government and military operations.
- The production type was restricted to new/original procurements or modifications/upgrades.
- Our analysis used only the lines containing direct unit counts for 2016–2026 (representing the Future Years Defense Program).

We discussed these restrictions with Janes representatives to ensure that these filters operated as intended and did not exclude relevant results.

Characterizing UxS DIB Elements

It is challenging to characterize the demand signal for UxS as a function of budgets and platforms that are being funded over some time horizon. However, we took it one step further by trying to characterize not just the platforms but also their components because demand on the UxS DIB might be best evaluated at the component level. It is useful to know what the potential

---

[9] AUVSI, *2021 Defense Budget for Unmanned Systems and Robotics*, Arlington, Va., 2021a.

[10] Janes, "Unrivalled Trusted Intelligence," webpage, undated; Janes, *Janes Markets Forecast*, brochure, Washington, D.C., 2020.

future demand for those components will be because this information can feed into an assessment of UxS DIB capacity.

Unfortunately, we found this equally if not more challenging because of the paucity of aggregate data on UxS components. Nevertheless, it is likely an area that OUSD(R&E) can continue to build on into the future.

UxS Platform Sample Sets

To build a clear representation of the demand signal for UxS, we developed a sample-set approach to create a snapshot of UxS procurement in the near term. The primary motivation for this approach was to help us manage the scope of this project, considering the immense volume and variety of platforms. The AUVSI UAS database identifies more than 4,000 separate platforms. Developing sample sets allowed us to narrow the focus of both our component-level analysis and our capacity analysis.

The goal of the sample-set approach was to create a list of systems across the following core dimensions:

- user (e.g., defense, commercial, recreational)
- use case (e.g., surveillance and reconnaissance, agricultural, explosive ordnance disposal)
- system type (e.g., UAS, UGS, USV, UUV)
- system size.

We endeavored to build a snapshot of the systems that are most indicative of each aspect of the larger UxS industry in each country of interest. We derived the size categories from U.S. government and military definitions in Federal Aviation Association, U.S. Navy, and U.S. Army documents, among others. We selected systems for each category in the sample sets based on U.S. government reports of UxS missions, news articles, and industry publications on UxS capabilities. We solicited input from RAND subject matter experts (SMEs) to ensure that we had collected data on relevant systems in each category. Given that we were also tasked with a near-peer assessment of China and Russia and an analysis of international UxS export markets, we also developed sample sets for those countries, as well as an international set—that is, UxS produced by countries other than China, Russia, or the United States. The size categories used and the platforms selected for the sample sets are presented in Appendix B.

Component Identification and Organization

We needed a framework to account for the components necessary for autonomy, as well as components shared by autonomous and nonautonomous UxS. We adopted the Work Breakdown Structure (WBS) as such a framework. Although a WBS is primarily a tool to aid cost estimators in developing relative order-of-magnitude estimates of platforms and weapon systems, it also provides a structure for listing components and subcomponents.[11] We developed WBSs for each type of UxS that focused on the vehicle components and payload sublevels. A simplified WBS is presented in Table 1.1.

Table 1.1. The Simplified UxS Work Breakdown Structure

Element Level	Level Description
1	UxS
1.1	UxS integration, assembly, test, and checkout
1.2	Platform/vehicle
1.2.2	Hull and structure
1.2.3	Propulsion
1.2.4	Energy storage/conversion
1.2.5	Electrical power
1.2.6	Vehicle C2—guidance, navigation, and control
1.2.7	Surveillance
1.2.8	Communications/identification
1.3	Payload 1 . . . *n* (specify)

NOTE: WBS level 1.1 does not constitute a component or payload; it is reserved for integration, assembly, test, and checkout by the prime integrator. It was not included in our framework but is shown here for completeness.

To identify components, we selected platforms from the sample sets and then investigated manufacturer data, government data (where relevant), and any other third-party information available to develop a list of components for each platform in the sample set. Ultimately, we concluded that information on platform components is highly variable, but, where possible, we used the information collected to inform our UxS DIB capacity assessment, as well as the questions we asked of interviewees.

Interviews

Our research approach involved a review of relevant literature (cited throughout this report) and information obtained through in-depth discussions with representatives and SMEs from key stakeholder organizations. The government entities were as follows:

- Air Force Global Hawk Program Office, Air Force Life Cycle Management Center/ RQ-4 Global Hawk, and Air Force Materiel Command
- Army Robotic and Autonomous Systems Program Office Program Executive Office Combat Support and Combat Service Support

---

[11] U.S. Department of Defense (DoD), *Department of Defense Standard Practice Work Breakdown Structures for Defense Materiel Items*, Washington, D.C., MIL-STD-881E, October 6, 2020b.

- Navy and Marine Corps Small Tactical Unmanned Aircraft Systems Program Office, Naval Air Systems Command/PMA-263
- Navy Unmanned Maritime Systems Program Office, Program Executive Office Unmanned and Small Combatants, Naval Sea Systems Command/PMS-406
- Navy Littoral Combat Ship Mission Modules Program Office, Program Executive Office Unmanned and Small Combatants, Naval Sea Systems Command/PMS-420
- DIU (autonomous systems).

Industry entities included the following:

- AUVSI
- Boeing Orca program
- Boeing RQ-21 Office
- General Dynamics Land Systems
- General Dynamics Mission Systems
- Huntington Ingalls Industries
- L3Harris
- Skydio
- Vayu Aerospace Corporation.

We developed a master list of questions for the interviews, which we adapted depending on whether the interviewees were government or industry representatives. All interviewees were provided with informed consent, and appropriate RAND human subjects protection practices were followed. The full interview protocol can be found in Appendix C.

Modified Fragility and Criticality Assessment

We used a modified fragility and criticality (FaC) approach to assess UxS DIB capacity. The original FaC methodology is described in a series of papers and briefing charts published between 2014 and 2016 by the Office of the Deputy Assistant Secretary of Defense for Manufacturing and Industrial Base Policy (ODASD[MIBP]).[12]

The FaC assessment process was devised and refined between 2011 and 2016 as a means of assessing DIB capabilities for the purpose of comparing them with defense requirements and prioritizing shortfalls for mitigation efforts. It is intended to be "rigorous, repeatable, transparent,"

---

[12] See, for example, Lirio Avilés and Sally Sleeper, "Identifying and Mitigating the Impact of the Budget Control Act on High Risk Sectors and Tiers of the Defense Industrial Base: Assessment Approach to Industrial Base Risks," in *Proceedings of the Thirteenth Annual Acquisition Research Symposium*: Vol. I, *Wednesday Sessions*, Monterey, Calif.: Naval Postgraduate School, April 30, 2016; and Sally Sleeper and John F. Starns, *Implementing Filters to Identify and Prioritize Industrial Base Risk: Rules of Thumb to Reduce Cognitive Overload*, Washington, D.C.: Office of the Deputy Assistant Secretary of Defense for Manufacturing and Industrial Base Policy, ADA623501, 2015.

and applicable across a wide variety of defense programs and industrial sectors.[13] The FaC approach consists of two distinct sets of assessment elements:

1. *Criticality* assesses the extent to which a capability (technology, part, or component, product) is difficult to replace if lost.
2. *Fragility* assesses the likelihood of that capability being lost.

In combination, criticality and fragility are a means of prioritizing resources to mitigate risk, where risk is defined as the likelihood and severity of negative outcomes; thus, FaC provides an indication of risk. ODASD(MIBP) experimented with alternative FaC factors through a pilot program,[14] and ultimately it settled on six criticality measures and four fragility measures, which are listed and described in Table 1.2.

Table 1.2. Revised Fragility and Criticality Factors for ODASD(MIBP) 2014 Assessments

Criticality: Characteristics that make a specific *capability* difficult to replace if disrupted (capability = technology, part, component, and product)	
Defense-unique capability	To what degree is the market for this capability commercial?
Skilled labor requirements for the capability	To what degree are specialized skills needed and available to integrate, manufacture, or maintain this capability?
Defense design requirements	To what degree is defense-specific knowledge required to reproduce this capability, an alternative, or the next-generation design?
Facility and equipment requirements	Are specialized equipment or facilities needed to integrate, manufacture, or maintain this capability?
Reconstitution time for the capability	What is the impact on DoD in terms of time to restore this capability if it is lost?
Availability of alternatives	To what degree are cost-, time-, and performance-effective alternatives available to meet DoD needs?
Fragility: Characteristics that make a specific *capability* likely to be disrupted (Will the department receive what it needs when it needs it?)	
Financial outlook (current provider)	What is the risk of this facility going out of business or exiting the market for this capability?
DoD sales (current provider)	How much total sales for this facility are from DoD contracts?
Firms in sector (existing market)	How many firms currently participate in this firm's market for this capability?
Foreign dependency (existing market)	What is the dependence on foreign sources for this capability?

SOURCE: Sleeper, Warner, and Starns, 2014, Table 5.

---

[13] Sally Sleeper, Gene Warner, and John Starns, "Identifying and Mitigating Industrial Base Risk for the DoD: Results of a Pilot Study," in *Proceedings of the Eleventh Annual Acquisition Research Symposium*: Vol. II, *Thursday Sessions*, Monterey, Calif.: Naval Postgraduate School, April 30, 2014.

[14] Sleeper, Warner, and Starns, 2014.

We identified FaC as a useful methodology for this project because it provides measures that have been found to be useful in practice across a wide range of defense programs. However, it was not a perfect tool for our purposes. The method's development and refinement were partly motivated by the Budget Control Act, which led to cuts; a loss of production capabilities; and a need to identify, prioritize, and mitigate the damage to existing (and, in many cases, legacy) programs. To this end, much of its focus is on identifying existing production assets that might be difficult to reacquire if production capability needed to be reconstituted in the future.[15]

In contrast, UxS involve rapidly evolving technology, demand, and supply factors. The inputs into production of UxS capability—and, particularly, autonomous capability—are still emerging. Our purpose is to project forward and evaluate the risk that those capabilities will not emerge, with additional emphasis on capability gaps relative to near-peer competitors. We determined that there are factors not incorporated in the original FaC that are important indicators of (and, in many cases, precursors for) developing those resources needed to provide the capability. Therefore, we modified FaC by changing one factor and expanding our assessment beyond FaC measures to incorporate consideration of enablers of emergence, expansion, and security of DoD access to UxS capabilities.

We removed reconstitution time since elements of the UxS DIB are not yet capabilities. Instead, we used a factor included in earlier versions of FaC—long lead time, which captures the question, "What is the impact on DoD from the lead time to obtain the capability from the current market?"[16] Given that UxS rely on novel technologies, some components or entire systems might have lengthy production processes, potentially including the transportation of raw materials or components. This suggests that it might be difficult to replace those parts or components if a source is lost. Given the preponderance of supply chain disruptions at the time of this research (a result of the COVID-19 pandemic), it is an important and timely criticality factor to assess. Furthermore, it is an appropriate and more direct substitute for reconstitution time while getting at the same question. Table 1.3 shows the factors used in our assessment and gives examples for each one.

In Chapter 3 we review the UxS DIB against these factors, but we considered other factors that could enable or inhibit the development of secure, reliable sources of supply for UxS technologies, such as the regulatory environment and need for product security.[17]

---

[15] Avilés and Sleeper, 2016.

[16] Sleeper, Warner, and Starns, 2014.

[17] Other categories that might be useful include public finance, the rate of technological change, and production expansion capacity.

Table 1.3. UxS Defense Industrial Base Fragility and Criticality Factors

Type	Name	Definition/Example
Criticality	Defense-unique capability	To what degree is the market for this capability commercial?
Criticality	Defense design requirements	To what degree is defense-specific knowledge required to reproduce this capability, an alternative, or a next-generation design?
Criticality	Skilled labor requirements	To what degree are specialized skills needed and available to integrate, manufacture, or maintain this capability?
Criticality	Facility and equipment requirements	Are specialized equipment or facilities needed to integrate, manufacture, or maintain this capability?
Criticality	Long-lead-time items	Are there materials, components, or other required items that take substantially longer to manufacture or obtain than other system inputs?
Criticality	Availability of alternatives	To what degree are cost-, time-, and performance-effective alternatives available to meet DoD needs?
Fragility	Financial outlook and demand certainty	What is the risk of this facility going out of business or exiting the market for this capability? What is the certainty of demand?
Fragility	DoD sales	How much in total sales for this facility are from DoD contracts?
Fragility	Number of firms in sector	How many firms currently participate in this firm's market for this capability?
Fragility	Foreign dependence	What is the dependence on foreign sources for this capability?

SOURCES: Sleeper, Warner, and Starns, 2014; authors' analyses of interviewee input.

The remainder of this report documents our demand signal analysis (Chapter 2), our analysis of the capacity of the U.S. DIB (Chapter 3), our equivalent analyses of China and Russia's UxS DIBs (Chapters 4 and 5, respectively), and our analysis of the international DIB (Chapter 6). Chapter 7 presents our conclusions and recommendations. This report also includes three appendixes that provide additional details on our analyses and approach: Appendix A provides data on UxS imports and exports, Appendix B documents the UxS sample sets, and Appendix C contains our interview protocol.

Chapter 2. Demand Signals

In this chapter we document the results of our demand signal analysis.

Association for Uncrewed Vehicle Systems International Budget Data Analysis

AUVSI provides analyses of DoD budget documents based on an understanding of the abbreviations and project lines in the budget documents, though its report is restricted to the current fiscal year.[1] AUVSI does not provide a comprehensive list of the budget lines in the DoD documents that its analysts use to build its data tables. Figure 2.1 shows AUVSI's breakdown of the UxS budget for FY 2021 by service and funding type.

Figure 2.1. Fiscal Year 2021 Department of Defense UxS Funding (in millions of dollars)

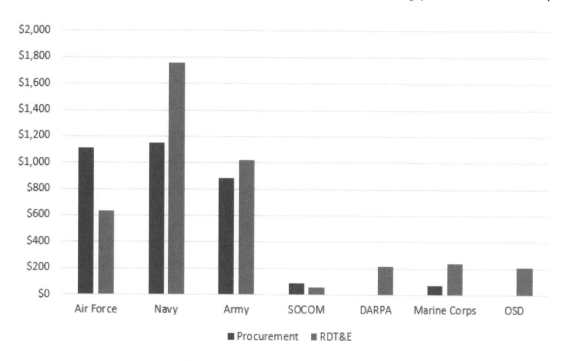

SOURCE: Authors' analysis of data from AUVSI, 2021a.
NOTE: OSD = Office of the Secretary of Defense; SOCOM = Special Operations Command.

According to the AUVSI data, the Air Force was more focused on procurement, the Navy was significantly more invested in RDT&E, and the Army was investing at similar levels in both.

---

[1] AUVSI, 2021a.

Comparably, DARPA, the OSD, and the U.S. Marine Corps (USMC) were focused primarily on RDT&E. Overall, RDT&E UxS expenditures in the FY 2021 budget were outpacing procurement, $4.1 billion to $3.3 billion, respectively.

When broken down by domain, we see that procurement across the services was heavily focused on the air domain, which accounted for 79 percent of DoD's UxS procurement budget for the year. When we look at platform types, we see significant investment in UAS, as shown in Figure 2.2.

Figure 2.2. Fiscal Year 2021 Department of Defense UxS Procurement, by Platform Type and Service (in millions of dollars)

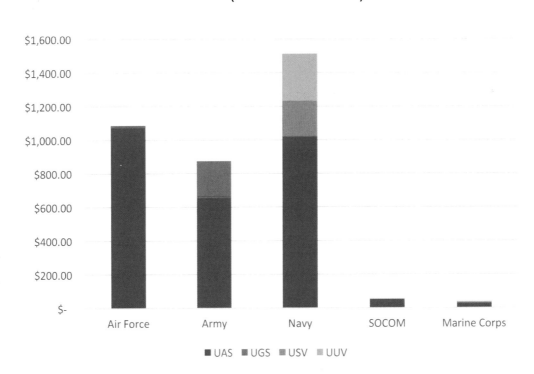

SOURCE: AUVSI, 2021a.

The services in FY 2021 spent a combined total of $2.8 billion on procurement of UAS, whereas they spent only $284 million on UUVs, $241 million on UGSs, and $210 million on USVs. In FY 2021, both the Army and the Navy spent more on procurement of UAS than on UxS in the ground and maritime domains, respectively.

As noted, AUVSI provides a snapshot of the current UxS budget lines; however, it does not provide any analysis of future procurement that can be used to characterize the demand signal. Furthermore, it does not report on the numbers of units being produced, nor is it clear how AUVSI determines which platforms to include in the uncrewed category and which to exclude. To obtain additional information on future procurement and the number of systems being produced, we procured the Janes Markets Forecast data set for UxS.

Janes Markets Forecast Data Analysis

The Janes data provide two things we were not able to obtain elsewhere: (1) a forecast of future procurement and (2) numbers of platforms being produced.[2] In this section we discuss our analysis of the Janes data.

Expected Units Delivered

When we separated the Janes data by system type and use, as shown in Figure 2.3, we found that annual U.S. UAS procurement (in numbers of units) was expected to increase by 161 percent over the period 2021–2026, along with a projected historic increase of 2,381 percent from 2016 to 2026.

Figure 2.3. U.S. Historical and Expected UxS Units Delivered per Year, 2016–2026

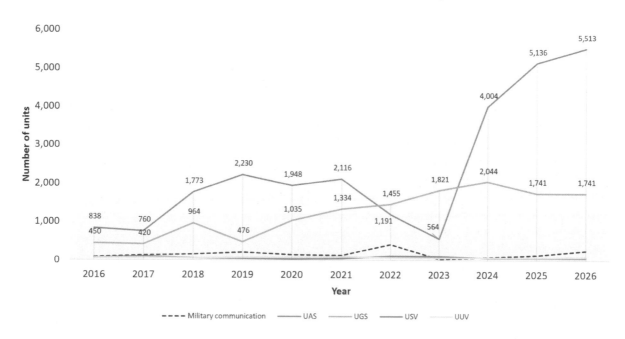

SOURCE: Authors' analysis of data from Janes Markets Forecast.
NOTE: The Military Communication category consists primarily of upgrades to legacy systems.

Notably, there was an expected 73 percent decrease from 2021 to 2023 before an 877 percent increase (from 564 to 5,513 units) in yearly procurement from 2023 to 2026. We found significant variability in the expected UAS units delivered (by unit) over 2021–2026 but concluded that the overall trend reflected increasing deliveries. Expected UGS deliveries were projected to increase 31 percent over the period 2021–2026, with demand peaking in 2024.

---

[2] Janes, "Janes Markets Forecast – UAV, UGV, USV Program Forecast," data as of October 27, 2021, and May 11, 2022.

USVs and UUVs followed a somewhat similar trend in that their expected delivery rate was expected to decrease from 2021 to 2023 before increasing again; however, their projected rate of increase into 2026 was significantly lower than for UAS and, in the case of UUVs, it was slightly negative. Because these changes are small relative to UAS and UGSs, we show the USV, UUV, and Military Communication categories in more detail in Figure 2.4.

Figure 2.4. U.S. Historical and Expected UxS Units Delivered, 2016–2026: UUVs, USVs, and Military Communication

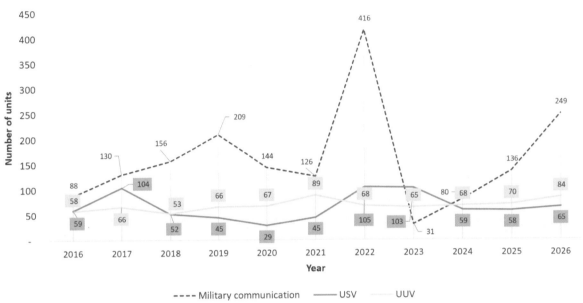

SOURCE: Authors' analysis of data from Janes Markets Forecast.
NOTES: The scale has changed from the previous figure. USV data point background is in green, and UUV in yellow. The Military Communication category consists primarily of upgrades to legacy systems.

Expected deliveries of USV units were predicted to increase 44 percent over the period 2021–2026.

UUV demand was projected to decline by 6 percent from 2021 to 2026 after decreasing 27 percent from 2021 to 2023 and then increasing by 29 percent from 2023 to 2026. We concluded that expected UUV demand would be variable but decrease slightly over the period from 2021 to 2026.

As we explored the data further, we found that the UAS demand was driven by rotary-wing platforms, as Figure 2.5 shows.

The rotary-wing category also includes small multicopters. Rotary-wing UAS procurement was predicted to rise from 1,324 units in 2021 to 4,385 units in 2026, while ground vehicle procurement was projected to rise slightly, from 1,357 units to 1,741 units, over the same period.[3] The top platforms (in numbers of units) were the UGS standoff robotic explosive

---

[3] Janes, "Janes Markets Forecast – UAV, UGV, USV Program Forecast," data as of October 27, 2021, and May 11, 2022.

Figure 2.5. U.S. Expected UxS Procurement, by Janes Platform Type, 2021–2026

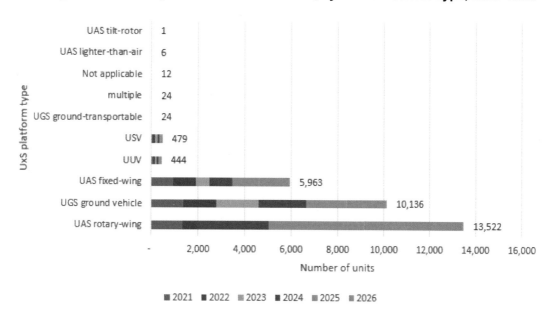

SOURCE: Authors' analysis of data from Janes Markets Forecast.

hazard detection system, with 3,190 units between 2021 and 2026, followed by three UAS: the short-range micro-UAS (SRM-UAS), with 3,000 units; the soldier borne sensor (SBS) micro-UAS, with 2,700 units; and Black Hornets, with 2,000 units. Across the entire unit count, most of the UxS are legacy systems. The cost per unit for the SRM-UAS was $34,000, and the SBS micro-UAS was $28,000.

Expected Production Costs

The Janes data included total historic and expected production costs, which were expected to increase between 2016 and 2026 for all platform types, with total UAS production costs increasing 130 percent, UGSs increasing 614 percent, USVs increasing 515 percent, and UUVs increasing 206 percent, as shown in Figure 2.6.

From 2021 to 2026, the total expected demand in terms of production costs (for the four types) was projected to grow from approximately $1.6 billion to $4.8 billion, an increase of 176 percent over the period. The FY 2021 $1.6 billion differed substantially from AUVSI's estimate of $3.3 billion. In Chapter 3 we explore what this demand increase—particularly for low-cost rotary-wing UAS—means for the DIB.

When we computed the average cost per unit, we found that the rotary-wing platforms were relatively lower in cost than fixed-wing platforms. Rotary-wing UAS cost an average of $1.1 million across all platforms, whereas fixed-wing UAS were significantly more expensive, approximately $10.3 million per unit, as Figure 2.7 shows.

Figure 2.6. U.S. Expected UxS Production Costs, by Janes Type and Use (in millions of dollars)

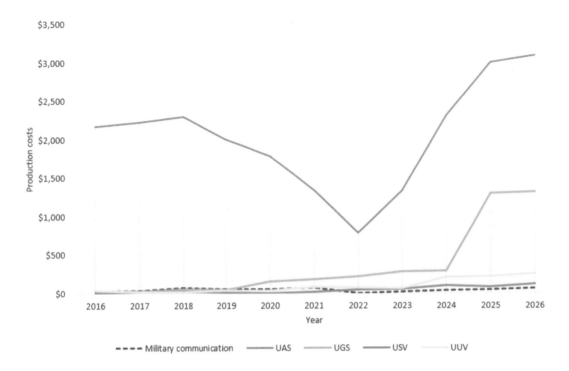

Figure 2.7. U.S. Average Unit Price of Expected UxS, by Janes Platform Type (in millions of dollars)

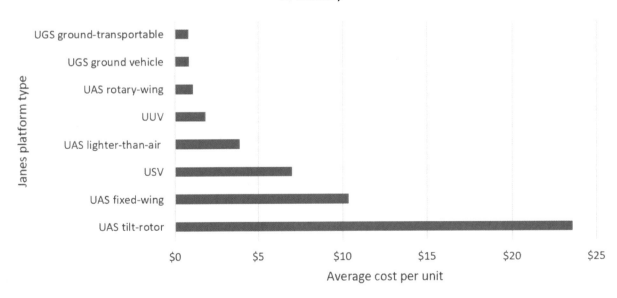

The distribution of cost per platform for the high-demand signal rotary-wing UAS (the SRM-UAS, the SBS micro-UAS, and the Black Hornet), as shown in Figure 2.5, are at a lower cost relative to legacy platforms, as shown in Figure 2.8.

Figure 2.8. Distribution of Expected Unit Cost for Rotary-Wing Platforms (in thousands of dollars)

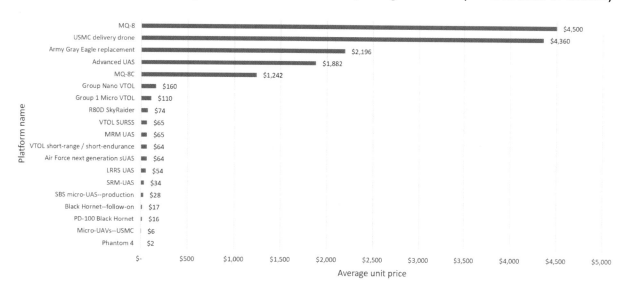

SOURCE: Authors' analysis of data from Janes Markets Forecast.
NOTE: LRRS = long-range reconnaissance and surveillance; MRM = medium-range mobile; SURSS = Small Unit Remote Scouting System; VTOL = vertical takeoff and landing.

SRM-UAS were priced at an average of $34,000 per unit; SBS micro-UAS averaged $28,000, and Black Hornets averaged $16,400. So, although we see a demand spike in numbers of UAS units (particularly for the period 2024–2026), the demand is driven by low-cost rotary-wing aircraft.

Limitations

Overall, the Janes data provided a longer time series of ongoing UxS procurement and funding by aggregating multiple sources and forecasting UxS procurement demand in the near term. However, the demand signal was still not complete as a result of incomplete open-source data. For example, some data records have unspecified platforms or unknown countries of manufacture. It is important to note that these data may not be comprehensive; in particular, they may not include smaller purchases, such as those that are not DoD programs of record.

Commercial Demand

In addition to our analysis of DoD demand, we sought to identify a commercial demand signal for UxS platforms to establish the relative sizes of the defense and commercial markets for UxS, as well as overall production capacity and potential competition for resources. We searched for this information by analyzing blogs, industry data, and Amazon lists. Overall, this information is difficult to obtain because companies do not regularly publish data on sales of individual products. Using publicly available data, we found sources estimating UAS sales at between 178,000 and 2.4 million units worldwide for 2023 across commercial drone companies.[4] Despite this wide range, even on the lowest extreme, the DoD demand signal is a fraction of the estimated commercial demand. We also encountered limited U.S. government data, but the Federal Aviation Administration (FAA) estimated that 10,000–70,000 model UAS and 36,000–180,000 nonmodel UAS were registered in 2022.[5] The Janes estimate of 1,200 units in 2022 is 1 to 3 percent of this range (1,200 of 250,000 and 46,000, respectively).

Assessments of the U.S. and global market for UxS are typically behind a paywall, and it is unclear how good they are. We used available free information from a variety of sources and extrapolated where growth rates are provided to estimate the size of the UAS market, as shown in Table 2.1.

Table 2.1. Summary of UAS Market Estimates for 2021 (in billions of dollars)

Source	U.S. or North America	Global
Fortune Business Insights[a]	$4	$14*
Precedence Research[b]		$9**
Allied Market Research[c]	$11**	$28**
Markets and Markets[d]		$24**
Industry Arc[e]		$5**

* Extrapolated linearly; **extrapolated based on source's compound annual growth rate.
[a] Fortune Business Insights, *Unmanned Aerial Vehicle (UAV) Market Size, Share & COVID-19 Impact Analysis*, Maharashtra, India, July 2020.
[b] Precedence Research, "UAV Drones Market Size to Worth Around US$ 102.38 Bn by 2030," March 1, 2022.
[c] Allied Market Research, "Unmanned Aerial Vehicle (UAV) Market by Type," webpage, October 2021.
[d] Markets and Markets, "Unmanned Aerial Vehicle (UAV) Market by Point of Sale, Systems, Platform (Civil & Commercial, and Defense & Government), Function, End Use, Application, Type, Mode of Operation, MTOW, Range, and Region (2021–2026)," June 2021.
[e] Industry Arc, "Unmanned Aircraft Systems (UAS) Market—Forecast (2022–2027)."

---

[4] For example, Statista forecasted that global shipments of commercial and industrial drones would increase from 5 million in 2020 to 7.1 million in 2025; see Statista, "Enterprise Drone Unit Shipments Worldwide from 2020 to 2030 (in Millions)," February 14, 2022. Gartner estimated that worldwide shipments would increase from 526,000 in 2020 to 1,273,000 in 2023; see Gartner, "Gartner Forecasts Global IoT Enterprise Drone Shipments to Grow 50% in 2020," press release, Egham, UK, December 4, 2019.

[5] FAA, *FAA Aerospace Forecast: Fiscal Years 2019–2039*, Washington, D.C., 2019.

Given that the Janes and AUVSI UAS procurement for 2021 are estimated at $1.4 billion and $2.8 billion, respectively, the few data points we have suggest that DoD represents anywhere from 13 percent to 70 percent of the North American market average, a large range of uncertainty. The fact that some global market estimates are lower than North American estimates also suggests that data surrounding these estimates are difficult to trust. Regardless, the lowest 13 percent figure indicates that DoD is a bigger player on a per-dollar level than it is on a per-unit level.

Summary of Findings

Our findings pointed to several overall conclusions about the demand signal by UxS type:

- Trends indicate that UxS unit demand is increasing, but only UAS demand was expected to increase more than 100 percent from FY 2021 levels. This effect is driven largely by demand for small, low-cost platforms from FY 2024 to FY 2026.
- The UGS demand increase was significant from FY 2019 to FY 2021 and was expected to increase 58 percent over the period from FY 2021 to FY 2026, with demand peaking in 2024 and at 460 percent in terms of total cost.
- Expected USV and UUV demand was expected to remain essentially flat from FY 2021 to FY 2026, but this still represents an increase over historical numbers.
- We saw a range of $1.6 billion (Janes) to $3.3 billion (AUVSI) in U.S. UxS procurement in FY 2021. The $1.6 billion estimate was projected to increase to $4.8 billion in FY 2026. If we extrapolate the $3.3 billion AUVSI estimate at the same growth rate, we obtain a projected increase to $8.4 billion in FY 2026.
- While DoD seems like a small player on a per-unit volume level (1–3 percent for UAS), it is a larger player on a per-dollar basis (13–70 percent of the North American UAS market); market data are difficult to trust.

U.S. UxS demand in the near term is focused on UAS across the military services, with the Navy spending the most on future long-term RDT&E. A large fraction of these systems are small rotary-wing aircraft, which aligns with recent statements by the Chief of the Office of Naval Research that "small and many" UxS platforms are suitable for short-term procurement due to their adaptability, low cost, and short production time.[6]

The rough demand signals for both DoD and the commercial sector show that DoD demand is very low compared with overall commercial demand. DoD demand in 2022 ranged from 1,200 to approximately 3,000 units (based on the trend line), suggesting that DoD UAS demand was roughly 1–3 percent of overall yearly U.S. UAS commercial demand.

---

[6] Duffie, 2021.

Chapter 3. U.S. Capacity Assessment

This chapter details our assessment of U.S. UxS production and sustainment capacity in the context of a modified FaC assessment. The objective of this assessment was to identify current U.S. capacity, estimate rough order-of-magnitude near-term (Future Years Defense Program) capacity as the market continues to develop, and assess the main risks for DoD.

To achieve these objectives, we drew on the FaC methodology described in Chapter 1, which highlights specific indicators, contributions, and risks to securing reliable sources of supply.

In this chapter we analyze the FaC factors using both qualitative and quantitative approaches. The qualitative analysis is largely derived from the interviews that we conducted with DoD officials and industry representatives, while the quantitative results come from an analysis of relevant data sources (as available). For FaC analyses, we first present findings that are common to all domains or types of UxS, followed by findings that are more specific to the air, maritime, and ground domains.

Modified Fragility Assessment

In this section we investigate fragility factors, starting with financial outlook.

Financial Outlook

Financial outlook refers to the risk of producers going out of business or exiting the market for UxS capabilities, as well as the certainty of demand.

All Domains

Demand Uncertainty

Representatives from six entities we interviewed, including two program offices and three prime contractors for defense UxS programs, identified demand uncertainty related to specific programs. They indicated that this uncertainty has led to a reduction in the number of firms in the supply chain, particularly for USVs and UUVs. (Note that we did find some evidence of a projected decline in demand for UUVs from 2021 through 2023 [see Figure 2.4], but we were not otherwise able to corroborate this claim.)

Interviewees indicated that uncertainty about future defense UxS demand deters firm investment in defense UxS programs, weakens supply chains, and amplifies workforce challenges. For instance, firms might invest in UxS by employing engineers and other personnel with UxS-specific skill sets who can help prepare proposals for future DoD UxS opportunities. They might also invest in facilities and equipment to meet the production rates anticipated for future contracts. Such investments might be an advantage in competitive UxS source selections and might benefit DoD by improving the technical quality and proposed production rates of submitted offers. But these investments are costly, and if DoD awards contracts less frequently or at lower values than the industry anticipated, firms might not recoup their investment. A pattern of inconsistent demand signals might increase firms' concerns that investments in UxS assets, such as skilled labor or production capacity, might not prove profitable.

Representatives from three UxS prime contractors and a UxS program office raised the concern that the more often DoD orders fewer platforms than initially anticipated, the less willing firms might be to make the up-front investments needed to respond to DoD's stated but uncertain requirements.[1] An interviewee from one UxS prime contractor provided the example of industry sometimes needing to pay to develop UxS technology to the point of demonstration before DoD will even announce a specific contract opportunity, and, even then, DoD might not follow a successful demonstration with a funded order. A representative of a second UxS prime contractor described having to substantially reduce production capacity for one DoD program after several consecutive years of defense orders coming in for lower quantities than the defense program initially planned. A representative from a DoD UxS program office provided an example of a UxS prime contractor investing in a cutting-edge manufacturing facility to meet anticipated DoD demand for a UxS platform when DoD had not yet committed to a production schedule.[2]

The challenge posed by demand uncertainty from one customer, such as DoD, can be mitigated by strong demand from other customers, such as those in the commercial sector. Revenue from these customers can offset lower-than-anticipated revenue from DoD, but only if the products demanded by alternative customers use the same production inputs, such that revenue from these other sources helps firms recover their fixed production costs and incentivizes maintenance of the production inputs needed to meet DoD orders. In addition, there are limitations on exporting defense technologies to foreign markets, so when too much of a supplier's product offerings are focused on DoD, the supplier could have the potential to diversify its customer base.

Prime contractors we interviewed stated that demand uncertainty made UxS defense work less appealing for them, but also made it difficult for them to maintain strong supply chains and attract and retain their UxS workforce.

Representatives from two prime contractors stated that small companies in their UxS component supply chains that do not have commercial customers struggle to cope with unexpected changes in demand from DoD. One interviewee pointed out that help managing volatile demand can be a good reason for small providers to work with larger prime contractors.[3]

Interviewees from three prime contractors stated that DoD demand uncertainty exacerbates labor problems: early career science, technology, engineering, and mathematics (STEM)

---

[1] Military UxS program official, interview with the authors, 2021; industry representatives, interviews with the authors, 2021.

[2] Military UxS program official, interview with the authors, 2021.

[3] Industry representatives, interview with the authors, 2021.

professionals leave the DIB for the commercial sector when defense programs are delayed. Given the benefits of defense-specific experience, this dynamic has lasting consequences, leaving defense firms struggling to retain workforces with defense experience. Similarly, slow defense technology refresh leads to cycles of innovation and production with intervening lags. Interviewees stated that it is hard to retain STEM and production-line professionals through long periods with no defense design or production work.[4]

Summarizing the challenges demand uncertainty presents, one interviewee pointed out that Congress has concerns about UxS technology, so it is quick to cut funding. But these funding cuts exacerbate many DIB problems and therefore increase the risks in developing and maintaining UxS technologies. In this way, congressional concerns become a self-fulfilling prophecy.[5]

Sample-Set Financial Outlook

We reviewed the financial state of the 20 first-tier suppliers in the U.S. sample set, as well as the second-tier suppliers of components that we were able to identify from our WBS categories. There were some duplicate companies, including separate subsidiaries of the same parent companies, but 11 out of 13 defense prime contractors are large or publicly traded with over $1 billion in annual revenues. The commercial primes are more likely to be small or privately held.

We examined commercial credit scores from Experian Information Services, Inc., to evaluate a selection of prime contractors and suppliers in the sample set.[6] This information is summarized in Table 3.1. Experian defines *financial stability risk* as the propensity for a company to go out of business in the coming 12 months, and *credit delinquency risk* as the propensity for a company to fall more than 90 days past due in its payment terms (to suppliers or other creditors) in the coming 12 months.[7] Nearly all the defense prime contractors were rated at low to medium risk for financial stability, except for Huntington Ingalls Industries, which was rated to be at medium-high risk. The credit delinquency risk for defense prime contractors also ranges from low to medium-high, with four of the 15 unique subsidiaries rated at medium-high risk (27 percent). Among the seven commercial primes, all were rated at low to medium risk for financial stability, but two were rated at medium-high or high risk for credit delinquency (29 percent).

---

[4] Industry representatives, interviews with the authors, 2021.

[5] Industry representative, interview with the authors, 2021.

[6] Experian Information Services, Inc., "Experian Commercial Credit Scores," data downloaded for selected companies through Nexis Uni, February 25, 2022, and March 8, 2022.

[7] Experian's financial stability risk scores are based on factors such as the percentage of delinquent commercial accounts, commercial derogatory public records, and risk associated with the industry sector and business type. Based on spot-checking, they appear to be consistent with credit ratings from other agencies. For example, Boeing is rated Baa2, Leidos is rated Baa3, and Huntington Ingalls is rated BBB–, whereas General Dynamics is rated A+. Ratings from other agencies are occasionally announced in the business press, but are not available to the public without a paid account.

Table 3.1. Experian Credit Scores for Selected Contractors

Company	Position	Type	System Name	Annual Revenue ($ millions)	Financial Stability		Credit Delinquency	
					Risk	Outlook	Risk	Outlook
Insitu (owned by Boeing)	Defense prime	UAS	ScanEagle, RQ-21	93,392	Low	Negative	Medium-high	Positive
Lockheed Martin	Defense prime	UAS	RQ-170	51,048	Medium	Negative	Medium	Positive
Northrop Grumman	Defense prime	UAS	RQ-4 Global Hawk	36,799	Medium	Neutral	Medium	Positive
Textron	Defense prime	UAS	Aersonde HQ	11,651	Medium	Negative	Medium-high	Positive
Draganfly	Commercial prime	UAS	Quantix Mapper V	4.4[a]	Medium	Neutral	Low	Positive
Freefly	Commercial prime	UAS	Alta X	N/A	Low	Neutral	Low	Positive
Impossible Aerospace	Commercial prime	UAS	US-1	34[b]	Medium	Positive	High	Positive
Skydio	Commercial prime	UAS	Skydio 2	N/A	Low	Positive	Medium	Positive
Foster Miller (owned by QinetiQ North America)	Defense prime	UGS	TALON	15,409	Medium	Negative	Medium-high	Positive
General Dynamics Land Systems	Defense prime	UGS	S-MET/ MUTT	37,925[c]	Low	Positive	Medium	Positive
Oshkosh Corporation	Defense prime	UGS	RCV L	6,857[c]	Low	Negative	Low-medium	Positive
Teledyne FLIR	Defense prime	UGS	Centaur, SUGV	3,086[c]	Low-medium	Positive	Medium-high	Positive
L3Harris	Defense prime	USV	Ghost Fleet Overlord	5,112	Medium	Positive	Medium	Positive
Leidos Holdings	Defense prime	USV	LUSV, Ghost Fleet Overlord	12,297	Medium	Negative	Medium	Positive
Vigor Industrial	Defense prime	USV	Sea Hunter II	678[d]	Low	Neutral	Medium	Positive
Saildrone	Commercial prime	USV	Surveyor	N/A	Low	Neutral	Low-medium	Positive
Sea Robotics	Commercial prime	USV	HYCAT	2.5	Low	Neutral	Low-medium	Positive
Bluefin Robotics	Defense prime	UUV	Bluefin 21	66	Low	Neutral	Low	Positive
Boeing	Defense prime	UUV	Echo Voyager	93,392	Medium	Neutral	Medium	Positive

Company	Position	Type	System Name	Annual Revenue ($ millions)	Financial Stability		Credit Delinquency	
					Risk	Outlook	Risk	Outlook
General Dynamics Advanced Information Systems	Defense prime	UUV	Black Pearl	37,925[c]	Low	Positive	Medium-high	Positive
Huntington Ingalls Industries	Defense prime	UUV	Orca	9,361[c]	Medium-high	Neutral	Low-medium	Positive
Phoenix International	Commercial prime	UUV	Artemis	3.8	Low-medium	Negative	Medium-high	Positive
Calculex	Supplier	UAS	RQ-4 Global Hawk	2.2	Low	Neutral	Low	Positive
Collins Aerospace	Supplier	UAS	RQ-4, RQ-21, ScanEagle	19,290[c]	Medium-high	Neutral	Low-medium	Positive
Cosworth AG	Supplier	UAS	RQ-21 Blackjack	81[c]	Medium	Neutral	High	Positive
Curtiss-Wright	Supplier	UAS	RQ-4 Global Hawk	2,391	Low	Positive	Medium	Positive
EaglePicher	Supplier	UAS	RQ-4 Global Hawk	N/A	Low	Negative	Medium-high	Positive
HoodTech Vision	Supplier	UAS	ScanEagle	N/A	Low	Positive	Low	Positive
Mercury Computer Systems	Supplier	UAS	RQ-4 Global Hawk	208	Medium-high	Negative	Medium-high	Positive
Rolls-Royce	Supplier	UAS	RQ-4 Global Hawk	15,726[c]	Medium-high	Neutral	High	Positive
Sierra Nevada Corporation	Supplier	UAS	ScanEagle	2,400[e]	Low-medium	Positive	Low	Positive
Trimble	Supplier	UAS	RQ-4 Global Hawk	3,148	Low-medium	Positive	Low	Positive
Vought (owned by Triumph Group)	Supplier	UAS	RQ-4 Global Hawk	1,870[c]	Medium-high	Neutral	Medium	Positive
Raytheon Technologies	Supplier	UAS, USV, UUV	RQ-4, Sea Hunter, Orca	56,587	Medium	Negative	Medium	Positive
B&G	Supplier	USV	Saildrone	8.8	Low	Neutral	Low	Positive
Kongsberg Maritime	Supplier	USV	Saildrone	2,817[c]	Medium	Neutral	Low	Positive
Valeport	Supplier	USV	Saildrone	9.6[c]	Medium-high	Neutral	Low-medium	Positive
Xylem	Supplier	USV	HYCAT	3,837	Low-medium	Neutral	Medium	Positive
Teledyne Marine	Supplier	USV, UUV	Knifefish, Saildrone	3,086[c]	Medium	Neutral	High	Positive
Beam Communications	Supplier	UUV	Orca, Knifefish	18[c]	Medium-high	Neutral	Low-medium	Positive
Edgetech	Supplier	UUV	Knifefish	2.6	Low	Positive	Low-medium	Positive

Company	Position	Type	System Name	Annual Revenue ($ millions)	Financial Stability		Credit Delinquency	
					Risk	Outlook	Risk	Outlook
Iridium	Supplier	UUV	Orca, Knifefish	583[c]	Low	Negative	Medium-high	Positive
Klein Marine Systems (owned by MIND Technology)	Supplier	UUV	Knifefish	21[f]	Low	Negative	Medium-high	Positive
Sonardyne	Supplier	UUV	Black Pearl	3.4	Low	Negative	Low-medium	Positive

SOURCE: Authors' analysis of Experian credit scores, 2022.
NOTE: In some cases, ratings of subsidiaries are different from those of the parent company. Revenues are listed for the parent company unless otherwise noted.
[a] Draganfly, "Draganfly Announces Record Revenue in Fourth Quarter and Fiscal 2020 Financial Results," press release, Vancouver, B.C., April 19, 2021.
[b] Parent company annual report, Alpine 4 Holdings, Inc., "Consolidated Statement of Operations," SEC Form 10-K for the fiscal year ended December 31, 2020, April 15, 2021.
[c] 2020 annual revenues were obtained from corporate annual reports. General Dynamics Corporation, SEC Form 10-K for the fiscal year ended December 31, 2021, February 9, 2022; Oshkosh Corporation, SEC Form 10-K for the fiscal year ended September 30, 2021, November 16, 2021; Teledyne Technologies, 2020 Annual Report, February 25, 2021; Huntington Ingalls Industries, 2020 Annual Report, February 11, 2021; Collins Aerospace revenues reported as a business segment in Raytheon Technologies Corporation, SEC Form 10-K for the fiscal year ended December 31, 2021, February 8, 2021; Global Database, "Financial Statements of Cosworth Group Holdings Limited," 2022; Rolls-Royce Holdings plc, Annual Report 2020, 2021; Triumph Group, 2021 Sustainability and Annual Report, 2021; Global Database, "Financial Statements of Valeport Limited," 2022; Beam Communications Holdings Limited, Annual Report for the Year Ending 20 June 2021, September 2021; and Iridium Communications Inc., 2020 Annual Report, April 2021;
[d] Reported annual revenues as of 2016, before the company was taken over by the Carlyle Group and Stellex Capital Management. See Andy Giegerich and Suzanne Stevens, "Vigor Industrial Sold to East Coast Private Equity Firms," *Portland Business Journal*, July 25, 2019.
[e] Estimated 2020 revenues; Forbes, "Sierra Nevada," webpage, undated.
[f] MIND Technology, Inc., SEC Form 10-K for the fiscal year ended January 31, 2021.

Nine of the 23 suppliers are large firms (or subsidiaries of large firms) with over $1 billion in sales. Suppliers were more likely than prime contractors to be rated medium-high risk for financial stability (six of 23, or 26 percent), and seven of the 23 (30 percent) were rated at medium-high or high risk for credit delinquency.

Findings

It is possible that significant uncertainty and changing demand signals from DoD has reduced capacity in the maritime sector of the DIB, particularly for larger USVs and UUVs. Interviewees also suggested that demand uncertainty deters investment in facilities and equipment and makes the industry less attractive to suppliers and skilled engineers. We were not able to corroborate these hypotheses, but quantifying the uncertainty in the demand signal and its broader effects on the industrial base is a potential area for future work.

Our review of FactSet data against first- and second-tier suppliers in the U.S. sample set was not useful for assessing the financial outlook of the sample-set firms, but publicly available information shows that most of the defense platforms we explored in our sample set were produced by large, financially healthy companies; only one of the defense primes was a smaller company with less than $100 million in annual revenues. Commercial credit ratings indicated that the suppliers we evaluated were somewhat less financially stable than the prime contractors.

DoD Sales

In this section we look at the extent to which firms currently participating in the UxS DIB are dependent on DoD sales. This metric provides insight into the extent to which firms may be exposed to risk should programs or research and development (R&D) projects get canceled or if funding is otherwise withdrawn. We explored market research data to supplement our analysis of this metric. However, since most of the defense primes are subsidiaries of large defense contractors, very little data are available on the extent to which their UxS sales are dependent on DoD.

All Domains

Having other customers can help firms ride out DIB demand uncertainty.[8] However, a low DoD share of sales can be a double-edged sword because the contractor is potentially less concerned about maintaining the capability to win DoD contracts and therefore less interested in riding out demand uncertainty. On the other hand, if the contractor is too dependent on DoD sales, it may not be able to ride out short-term reductions in production due to demand uncertainty.

The Air Domain

In the air domain, an interesting variation on the problem arose. Small UAS have low profit margins because there is a large commercial market with many potential suppliers. However, small orders from DoD are not necessarily attractive, especially if commercial off-the-shelf (COTS) systems need to be modified, making a small production run of a specialized system less profitable.[9]

The Maritime Domain

In the maritime domain, DoD is the majority buyer of large platforms and appears to dominate the demand for autonomy.[10] Stakeholders described growing commercial demand for USVs. Should this demand materialize, it could result in a smaller market share for DoD while increasing the risks highlighted earlier.[11] One UUV vendor reported that DoD was by far its largest customer and that remaining customers were foreign militaries.

---

[8] Industry representative, interview with the authors, 2021.

[9] Industry representative, interview with the authors, 2021.

[10] SME, interview with the authors, 2021.

[11] Industry representatives, interviews with the authors, 2021.

The Ground Domain

Stakeholders in the ground domain indicated that there were potential commercial markets for UGS capabilities, though not necessarily for the systems currently being developed.[12]

Findings

Unfortunately, data in this area are limited, but our interviews suggested that smaller platforms had more diversity in sales, whereas large platforms tended to be more dependent on DoD sales. This is an area for further research.

Number of Firms in the Sector

In this section we explore the number of firms in the various UxS markets. We used AUVSI data to establish a range of firms at the first tier and gained perspectives on these findings from our interviews. While we do not attempt to determine what a healthy number of firms is per market or platform type, we do assert that given the level of financial health for firms in a sector, which is a separate FaC dimension, more firms in a market is an indicator of lower risk. One justification for this assertion is that if one firm faces a disruption, DoD has a larger pool of potential alternatives to turn to. Further, we distinguish between the number of firms for platforms sold to defense-only and mixed-use markets since defense-unique platforms might require specialized production capability reducing the number of firms capable of replacing an existing firm.

All Domains

Table 3.2 shows the count of firms by market use and platform type.

Table 3.2. Count of U.S. Firms, by Market Use and Platform Type

Market Use	UAS	UGS	USV	UUV
Defense only	40	15	2	2
Mixed use (including defense)	108	28	14	30
Mixed use (excluding defense)	153	71	10	33
Total (after removing double-counted firms)	247	101	22	55

SOURCE: Authors' analysis of data from AUVSI, undated a.
NOTES: We excluded inactive platforms and any platforms with insufficient data. The row labeled Total lists the number of distinct firms in each domain (air, ground, maritime surface, and maritime underwater) across all market uses, because the same firm might produce platforms for different market uses. Therefore, the total in that row is less than the sum of the number of suppliers in all market uses. There are several firms offering both defense-only platforms and mixed-use (including defense) platforms. The number of unique firms offering defense-only platforms, mixed-use (including defense) platforms, or both are, by domain, 129 (UAS), 37 (UGS), 15 (USV), and 32 (UUV). Put differently, these are the number of unique firms producing any platforms in the first two rows.

---

[12] Military UxS program official, interviews with the authors, 2021.

29

We see that the number of manufacturers with platforms marketed specifically for defense purposes usually make up a small percentage of the overall market: 16 percent for UAS, 15 percent UGSs, 9 percent for USVs, and just 4 percent for UUVs. If we consider the number of manufacturers with a platform for the defense and mixed-use defense markets, the share rises appreciably to 52 percent of UAS, 37 percent of UGSs, 68 percent of USVs, and 58 percent of UUVs. In Figure 3.1 we explore the breakout further by the size of the platform. (See Appendix B for the size categories used.)

Figure 3.1. Number of U.S. Platform Manufacturers, by Type and Size

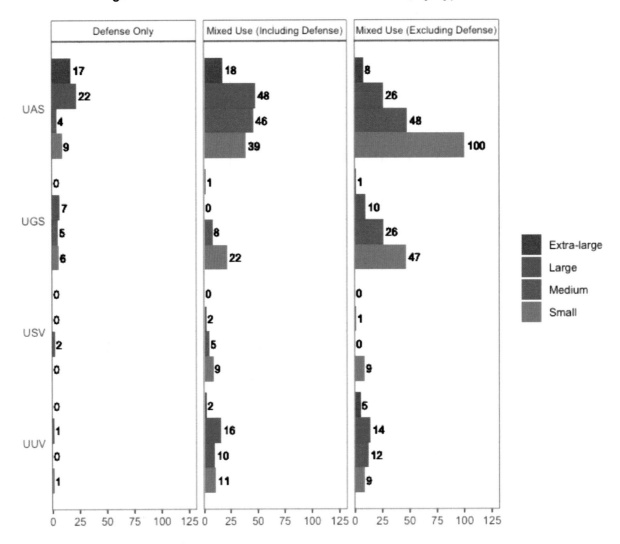

SOURCE: Authors' analysis of data from AUVSI.
NOTE: These data may include the same firm multiple times across size categories.

We can draw a few additional conclusions about the number of firms from Figure 3.1. First, defense-only platforms tend to be larger for UAS, a stark contrast to the mixed-use, excluding defense, market. In fact, this is a useful quantification in that these data show that the UAS manufacturer level is likely quite healthy.

We see very few manufacturers of large UGSs, which is a capacity concern. Similarly, we see few to no manufacturers of USVs except the small USV market, and a slightly healthier UUV manufacturer base.

We also explore the metric of the number of firms by expected 2026 budget, as shown in Table 3.3. It is unclear what a healthy dollars-per-firm value is, but it provides insight into one platform type versus another.

Table 3.3. Millions of Dollars per Firm, by 2026 Expected Budget

Type	Dollars per Defense-Only Firm ($M)	Dollars per Defense and Mixed-Use (Including Defense) Firm ($M)
UAS	77.5	24.0
UGS	95.3	38.6
USV	58	7.7
UUV	128	8.0

SOURCES: AUVSI, undated a; authors' analysis of data from Janes Markets Forecast.
NOTES: The expected 2026 budget is $3.1 billion for UAS, $1.3 billion for UGSs, $116 million for USVs, and $256 million for UUVs. Firms with defense only and defense and mixed-use (including defense) platforms are counted only in the Dollars per Defense and Mixed-Use (including Defense) Firm category.

The USV and UUV markets are highly uncertain. When viewed as a defense market only, the UUV market seems fragile relative to the future demand signal, whereas the USV market seems more stable relative to UAS and UGSs. Maritime firms noted that their volumes are low, which is confirmed by the number of dollars per defense and mixed-use firm. Given that there are only two USV and UUV defense-only firms, it is more likely this market is driven by vendors also in the commercial space, and given that the data show that there are enough firms, the problem in the future may be not enough demand.

The Air Domain

In the air domain, the number of suppliers depends on whether or not the UxS is a legacy system. If a platform is obsolete or approaching obsolescence, like the RQ-4, for example, there could be few or no options to source components. There are already few domestic suppliers of radios, controllers, and air vehicles and, as a system ages, these suppliers are not likely to retain capacity as new demand arises from newer systems that are likely to deliver large orders. Interviewees believed that, at the first tier, there are many options for small UAS, but there are far fewer for large or extra-large UAS—and this was something that our quantitative analysis of AUVSI data confirmed.[13]

---

[13] Industry representatives, interview with the authors, 2021; military UxS program official, interview with the authors, 2021.

The Maritime Domain

Stakeholders in the maritime domain indicated that volumes were too low to support multiple vendors at the prime and supplier levels, except where there were commercial analogues or other defense programs to provide additional revenue.[14] For example, there are many vendor options for watercraft that use components with commercial demand, such as engine and propulsion systems or sensors. For components with less commercial demand, like magnetic suite cables, fewer options are available. But again, when military or operational requirements dictate higher levels of accuracy—cybersecurity, for example—those same requirements reduce the number of qualified suppliers available.[15]

The Ground Domain

Interviewees indicated that more accuracy-dependent components, like navigation and related autonomy-enabling sensors, along with their accompanying software, have few or no alternative suppliers. There is an indication that there is also a firm capacity problem for smaller UGSs, as they have no alternative suppliers for critical components.[16] This is a bit surprising, but it seems to be corroborated by the AUVSI data signaling a small number of UGS platform manufacturers.

Findings

Overall, we see lower risk for large and small UAS platforms. We see high risk for larger UGSs. USVs and UUVs are a mixed picture, as defense-only UUVs seem more fragile, but both USVs and UUVs seem to have lower fragility risk when mixed-use platforms are considered.

Foreign Dependence

Finally, we reviewed sources of foreign dependence for UxS capabilities. We reviewed the components that were sourced not just from near-peer competitors but also allies and partners. When possible, we reference where in the supply chain these conditions are exhibited.

All Domains

The current supply chains for all DoD UxS platforms are global. In some cases, this is because UxS prime contractors have deemed a foreign source most competitive in terms of either price or quality. In other cases, there are no U.S. sources available, but there might be sourcing options spanning multiple countries. In some important cases, the only sources are in high-risk

---

[14] Industry representatives, interview with the authors, 2021.

[15] Industry representatives, interview with the authors, 2021; military UxS program official, interview with the authors, 2021.

[16] Military UxS program official, interviews with the authors, 2021.

countries, such as China. Dependence on these high-risk countries raises concerns about politically induced embargoes and possible subtle sabotage. Furthermore, mergers and acquisitions can create immediate dependencies that were otherwise unanticipated or even unknown, as was the case when a Chinese state-owned company bought an Italian UAS manufacturer.[17]

We identified foreign sources for a variety of components and materials: UAS motors, electro-optical sensor lenses and components, light detection and ranging (LiDAR) sensors, lithium-ion batteries, semiconductors, inertial navigation systems (INSs), winches, ferrite, and fiberglass.

Table 3.4 lists UxS components identified through our analysis and by interviewees as having foreign sources.[18]

Table 3.4. Foreign Sources of Platforms or Components

Components and Materials with Identified Foreign Sources	Domain	Originating Country	Sources Available in Other Countries?
Motors	UAS	Taiwan	Yes
Electro-optical sensor lenses and charge-coupled devices	UAS	Japan	Yes
LiDAR sensors	UAS	Switzerland	Unknown
Batteries	UAS, UGS, USV, UUV	China, Japan, South Korea	Unknown
Semiconductors	UAS, USV, UUV	South Korea, Taiwan	Unknown
INSs	USV, UUV	France	Yes
Winches	USV, UUV	Denmark, Sweden	Yes
Raw materials (ferrite, metals, fiberglass, some plastics)	UGS	Non-U.S countries	Unknown

SOURCE: Stakeholder and SME interviews with the authors, along with other corroborating sources.

One UAS motor is currently sourced in Taiwan, though alternative sources are available in the United States and elsewhere. Similarly, electro-optical sensors, INSs, and winches were identified as being sourced from specific foreign countries, despite the availability of alternative sources (albeit possibly with lower quality or at higher prices). Interviewees noted that microelectronics and raw materials that we discuss in the context of long lead times later in

---

[17] James Marson and Giovanni Legorano, "China Bought Italian Military-Drone Maker Without Authorities' Knowledge," *Wall Street Journal*, November 15, 2021.

[18] To protect potentially proprietary information on firms' supply chains for specific products, we avoid identifying components, materials, and platforms with enough detail to identify firms.

this chapter also have limited global sources. The following sections provide more detail on a selection of the components and materials discussed here.

Electro-Optical Sensor Components

Interviewees identified electro-optical sensor lenses and charge-coupled devices as products for which quality varies substantially by source. They named China, Japan, and, for some devices, Taiwan as countries with sufficiently high-quality products to meet U.S. military requirements.

Raw Materials

Raw or partially refined materiel for uncrewed platforms has foreign dependencies that were highlighted during the COVID-19 pandemic, though these dependencies existed before the pandemic.[19]

One example is ferrite, a magnetic material composed of iron oxide blended with small amounts of additional metallic elements that is found in some electric motors. China is currently the largest producer and consumer of ferrite magnetic cores used in motors. European countries produce high-end ferrite cores with starting material purchased from China.[20]

With increased demand for clean energy technologies, the U.S. government will be competing with the larger economy for these magnetic materials—along with, for example, lithium-ion batteries.

Composites

Fiberglass is a composite material of a fiberglass cloth and a vinyl ester polymer.[21] The United States produces just a fraction of the world's fiberglass cloth and has become reliant on material from China. The COVID-19 pandemic and a 2021 cold-weather natural disaster in Texas resulted in shortages of the fiberglass materials necessary for the bodies of UAS.[22]

There are several suppliers of high-end carbon fiber composites in the UAS market, yet manufacturers have reported difficulty acquiring carbon fiber, which has led to longer lead times.[23]

---

[19] Richard Silberglitt, James T. Bartis, Brian G. Chow, David L. An, and Kyle Brady, *Critical Materials: Present Danger to U.S. Manufacturing*, Santa Monica, Calif.: RAND Corporation, RR-133-NIC, 2013.

[20] Bodo Arlt, "Supply Chain Challenges of the Ferrite Industry as a Result of the Pandemic," *Bodo's Power Systems*, August 2021.

[21] Fiberglass cloth is a woven fabric made of glass; the cloth is joined with a resin to form fiberglass.

[22] Brett Haensel, "Surfers, Swimmers, Boaters Run Into Summer-Disrupting Fiberglass Shortage," Bloomberg, August 30, 2021.

[23] Hannah Mason, "Resin Shortages Continue to Affect Composites Supply Chain," *CompositesWorld*, May 3, 2021.

Lithium-Ion Batteries

Lithium-ion batteries are sourced largely outside the United States. At the top of the supply chain, raw materials, such as lithium salt brines and ore, are found and processed elsewhere by a limited number of suppliers. The U.S. Geological Survey reported that, as of 2020, "five mineral operations in Australia, two brine operations each in Argentina and Chile, and two brine and one mineral operation in China accounted for the majority of world lithium production." However, the processing of these raw materials into battery-grade chemicals is largely limited to China. Consequently, "lithium supply security has become a top priority for technology companies in the United States and Asia."[24]

In the middle of the lithium-ion battery supply chain, battery manufacturers themselves are concentrated in Asia. Several interviewees mentioned concerns about foreign dependency. They cited a lack of domestic manufacturers of batteries and battery components and said that they sourced their lithium-ion batteries from integrators or vendors who, in turn, sourced those batteries from Chinese firms. Others were reliant on batteries from Japan.[25] Although it is less likely that downstream companies that depend on Japanese lithium-ion batteries would be exposed to risk, this nonetheless represents a foreign dependency.

Competing demand from the growing electric vehicle market might be stimulating the expansion of mining and processing capacity, along with battery capacity; however, in the near term, the lithium market is experiencing shortages and high prices.[26]

An oft-discussed opportunity is the development of domestic lithium-ion battery supply chains that would decrease reliance on foreign sources while also reducing the constraints from limited suppliers within the United States, but that is not without its challenges. Should orders increase in the near term before new raw material sources develop, the prices of battery packs and related components could increase the costs of procuring these systems. Depending on the size and energy demand of a UxS, the cost of a battery pack could represent a large percentage of its total cost. For example, in electric vehicles, such as the Tesla Model III, battery packs represent 15–19 percent of the selling price.[27] These costs could increase the price of UUVs, which are more likely to depend on electric propulsion than UAS, UGSs, and USVs. Smaller UAS are also more likely to depend on electric motors and batteries and could incur higher costs.

---

[24] Brian W. Jaskula, *Lithium*, Washington, D.C.: U.S. Geological Survey, Mineral Commodity Summaries, January 2021, n.p. [p. 2].

[25] Industry representatives, interviews with the authors, 2021; military UxS program officials, interviews with the authors, 2021.

[26] This was due, in part, to ongoing pandemic-related supply chain issues, but it was largely a result of raw material producers cutting back supply while the battery and electric vehicle market experienced an increase in capacity. See Benchmark Mineral Intelligence, "Benchmark's Lithium Carbonate Prices Reach New All-Time Highs," press release, London, October 6, 2021.

[27] This calculation assumes a 70-kWh battery pack; see Trefis Team, "Are Battery Cost Improvements Still a Big Driver of Tesla's Margins?" *Forbes*, December 1, 2021.

Findings

We see foreign dependence in a variety of areas: UAS motors; electro-optical sensor lenses and charge-coupled devices; LiDAR sensors; batteries; semiconductors; INSs; winches; and such raw materials as ferrite, metals, fiberglass, and some plastics. We were not able to corroborate all of these. Batteries are perhaps the most significant dependence, but we also observed high levels of foreign dependence for other components, such as UAS motors and electro-optical sensor components.

Additional Factors

In addition to the fragility factors addressed so far, we considered regulatory issues, the defense acquisition process, technical standards, and legacy systems.

Regulatory Concerns

Autonomy and Cybersecurity

Autonomy and cybersecurity are both areas of regulatory ambiguity affecting market demand. We found that defense policies, processes, and acceptance criteria for autonomous systems are unclear, creating risk aversion and adding expense to UxS development.

For example, it is expensive and complex for companies that produce larger UAS to meet FAA requirements for airworthiness certification, and these requirements can result in more complex systems.[28] However, most users want systems that are simple to use and maintain. This division in customer priorities is a major driver of the high DoD market share for larger UAS. One potential reason there is currently no commercial market for large UAS platforms may be that regulations simply do not allow for it. In this regard, the fact that DoD testing for defense platforms is more clearly defined than that for commercial platforms might make the defense market attractive to potential primes and suppliers.[29] This is an area for further investigation.

This is not to say that DoD's testing procedures and regulations are simple or easy to navigate. Stakeholders reported that commercial vendors hesitate to get into the defense market because they perceive cybersecurity rules as complex and constantly changing.[30] Changing regulations also create disincentives that might render worthless huge capital investments made by industry.[31]

---

[28] FAA airworthiness certification is required for UAS that weigh more than 55 lbs and that have the capabilities to perform automated fleet operations, fly beyond visual line of sight, or sustained flight over populated areas. See, for example, Pilot Institute, "Airworthiness Certification of Drones—Everything You Need to Know," July 18, 2021.

[29] Industry representatives, interview with the authors, 2021; military UxS program officials, interview with the authors, 2021; DIU officials, interview with the authors, 2021; SMEs, interview with the authors, 2021.

[30] SME, interview with the authors, 2021.

[31] DIU official, interview with the authors, 2021.

The risk-averse nature of the testing community was noted as a roadblock to qualifying autonomous systems. For example, large UAVs are treated as aircraft in a standard pilot-copilot model, even though they contain no pilot. This makes meeting safety policies for autonomy a challenge.[32] Institutional and cultural friction could also present challenges. Interviewees noted that any reduction in the number of pilots to advance autonomy is an uphill battle against a culture that favors human pilots.[33] For USVs and UUVs, there are not yet clear policies, processes, procedures, and acceptance criteria for autonomous systems. Some safety standards for crewed sea vessels are clearly unnecessary for uncrewed vessels, yet they remain. These rules affect ship design, materials, and sea certification. These challenges are compounded when every redesign triggers a major review. UGS program and supplier stakeholders noted this risk aversion. Commercial demand can compensate for uncertain defense demand and, in some cases, increase the number of supplier options.[34] Currently, the maritime industrial base is limited primarily to nonautomated craft and engines for USVs and UUVs. To spur more commercial demand, regulatory bodies need to resolve their requirements and disseminate the criteria. For example, the International Maritime Organization, which publishes the *International Regulations for Preventing Collisions at Sea*, is determining the criteria for autonomous maritime systems.[35] Once those criteria are determined, suppliers can navigate them to meet requirements from DoD as well as for commercial seas lanes.

Increasing Domestic Sources

An additional challenge pointed out by several interviewees, representing both program offices and contractors, was a failure to enforce requirements. This failure asymmetrically disadvantages firms that invest the most in meeting DoD's requirements. One example given was the prohibition on Chinese sources under Section 848 of the NDAA for FY 2020.[36] As demonstrated by our examination of lithium-ion batteries, this policy could be enforced at one level of the supply chain (i.e., the battery pack) while reliance on Chinese sources for subcomponents or raw materials (i.e., the processing of battery-grade minerals) remains.

In addition, uncertainty arises when guidance is issued through the NDAA, which is a less-permanent means of conveying these types of rules. Firms can be uncertain as to how much to invest because policies could change from year to year.

---

[32] DIU official, interview with the authors, 2021.

[33] DIU official, interview with the authors, 2021.

[34] Industry representative, interview with the authors, 2021; military UxS program officials, interviews with the authors, 2021.

[35] SME, interview with the authors, 2021.

[36] Industry representatives, interview with the authors, 2021; DIU officials, interview with the authors, 2021.

The Defense Acquisition Process

DoD requirements, acquisition workflows, and employment practices can slow technology adoption. The takeaway we heard in our interviews was that the defense acquisition process is still too slow to avoid obsolescence.

Strong requirements for the safety and robustness of autonomous systems are needed to ensure their readiness for operation; however, the defense acquisition process slows DoD's adoption of new technology to the point that, according to one interviewee, UxS in development are at high risk of becoming obsolete before a prototype can be delivered. In some cases, it can result in suppliers outright refusing to work with DoD customers. For instance, one firm could not get a paint manufacturer to work with it because of the complexity of the government acquisitions process.[37]

Issues begin with the requirements process. Capability gaps must be identified to justify funding. Then, to fill the gap cost effectively, there should be demand from a major U.S. military command that is willing to fund any development efforts.

Acquisition pathways present the next hurdles. It is often necessary to engage multiple funding sources and to troubleshoot resource allocation processes and decisionmaking pathways. One program office reported that the traditional defense procurement process determines requirements up front, based on relatively mature technology, then moves into production.[38] Sometimes substantial investments in technology risk reduction are needed, extending timelines.

This approach is not compatible with fast-moving technology development and deployment cycles, especially for software- and information-heavy systems. To take advantage of the latest technologies, DoD needs to iterate quickly, get systems fielded, and then improve the capabilities and build the requisite manufacturing capacity for each system.[39]

UGS, USV, and UUV program stakeholders reported that even with middle-tier authorities, the defense acquisition process is too slow and burdensome.[40] Even faster pathways, like Joint Urgent Operational Needs, have suffered from slowdowns because they, too, need technical requirements specified up front.

A commercial entity that decides to respond to a DoD request for proposals will need to make many decisions about risks and trade-offs. In a fast-moving industry, offerers might need to invest enormous amounts of capital up front in responding to such a request. These companies assume the risk that their proposed solution will be obsolete by the time DoD makes a procurement decision, and thus any orders to recoup the cost of their investments could be lost.

---

[37] Industry representative, interview with the authors, 2021.

[38] Military UxS program official, interview with the authors, 2021.

[39] Military UxS program official, interview with the authors, 2021.

[40] Industry representatives, interview with the authors, 2021; military UxS program officials, interviews with the authors, 2021.

In many cases, primes and subcontractors need to work through solutions sequentially, but if demand does not materialize, that experience is lost to the DIB.[41]

The defense acquisition process also poses challenges to prototypes.[42] Beyond the expected rounds of design iteration, some projects and programs need to get prototypes into the hands of warfighters who can use them, report feedback, and iterate further still.[43] These multiple stages of assessment and testing all play a role in acquisition process slowdowns.

If a system is developed commercially and driven by nondefense needs, DoD-specific design requirements are likely to have been an afterthought. When DoD does realize the military value of an existing commercial technology, it is a tough sell for companies to revisit design aspects that run afoul of cybersecurity or other defense requirements.[44]

Finally, the traditional acquisition system creates risk aversion on the operational side. When commanders do receive these assets, as multiple interviewees pointed out, they can be quite conservative in using them. The long procurement time, along with limited funds, increases the pressure not to lose any units, prototype or otherwise.[45] Additional challenges for autonomous systems relate to safety, and operators could experience blowback should something go wrong.

All of these factors—requirements processes, acquisition pathways, options to harness other DoD resources, and institutional incentives for operational use work against the fast fielding of new and emerging capabilities.

To address some of these challenges, DoD can get involved in earlier stages of technology maturity rather than waiting for technologies to mature.

It is part of the DIU's role to reduce trade-offs for U.S. companies and DoD programs alike. It has the expertise to help firms understand the defense operating environment and special requirements, as well as vet firms for cybersecurity and security requirements.

Technical Standards

Representatives from several firms mentioned uncertainty about technical standards as a significant challenge. When DoD issues new cybersecurity standards or requirements for supply chains after contractors invest their own funds in developing technologies, prototypes, or production capability, it undermines the value of investments; deters participation in the DIB; and compromises speed, quality, and cost.

---

[41] Industry representative, interview with the authors, 2021.

[42] Industry representative, interview with the authors, 2021.

[43] SME, interview with the authors, 2021.

[44] DIU official, interview with the authors, 2021.

[45] SMEs, interview with the authors, 2021; military UxS program official, interview with the authors, 2021.

Legacy Programs

Systems tailored to defense uses suffer more from legacy system challenges than those marketed to the commercial sector. One example from our interviews was the RQ-4 program. The system was said to be so old that there were few suppliers of replacement parts. This had led the program to seek out any available producers, even if it meant foreign dependence.[46] One reason is that technology evolves more quickly in commercial markets than in the defense market, much faster than DoD and its primes can iterate.[47] Consequently, DoD must source commercially obsolete products, which comprise a very high share of business for these products. Keeping a company from shutting down an existing product line can also be difficult. Companies must invest resources in personnel, facilities, and equipment to be able to produce a component. When that component accounts for only a small share of a company's revenue, it is difficult to justify keeping the production line open.

Modified Criticality Assessment

Defense-Unique Capabilities

For this factor, we attempted to ascertain the degree to which the market for UxS capabilities is commercial. We looked at the number of different platforms by size, use, and type from the AUVSI database, as shown in Table 3.5, and collected insights through our interviews.

Table 3.5. Count of U.S. Platforms, by Market Use and Platform Type

Market Use	UAS	UGS	USV	UUV	Total
Defense only	84	21	3	2	109
Mixed use (including defense)	264	75	23	70	431
Mixed use (excluding defense)	299	142	20	111	571
Total	644	238	46	183	1,111

SOURCE: Authors' analysis of AUVSI, undated a.
NOTES: We excluded inactive platforms and any platforms with insufficient data. We included maritime semisubmersibles in the UUV category.

By this measure, the overall share of defense-unique platforms was 10 percent (109 out of 1,111 platforms); or 13 percent for UAS, 9 percent for UGSs, 7 percent for USVs, and 1 percent for UUVs. This is just one way to represent defense uniqueness, but it indicates that the market has many commercial options. Exploring the breakout of the platform market use by size is also interesting. Figure 3.2 shows that larger platforms tend to be more defense oriented, whereas smaller platforms tend to have more diversity and use in nondefense applications. There are few

---

[46] Military UxS program official, interview with the authors, 2021.

[47] DIU officials, interviews with the authors, 2021.

Figure 3.2. Composition of U.S. Platforms, by Type and Size

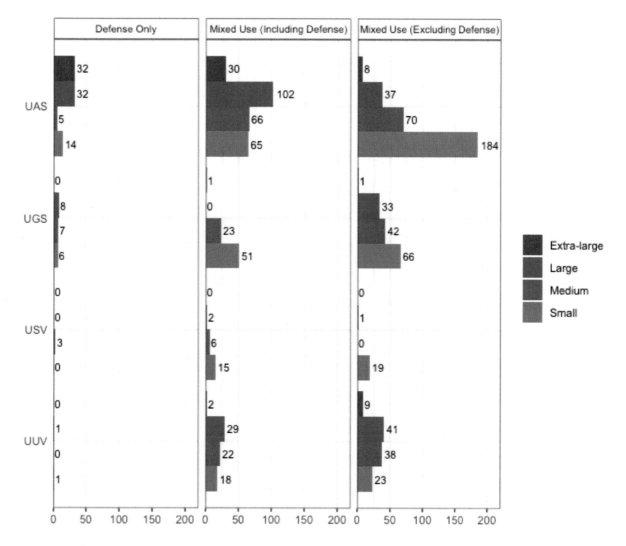

SOURCE: Authors' analysis of data from AUVSI.

platforms in the defense-only USV and UUV markets, but the UUV market seems to be a bit of an outlier in that there are more large-size platforms in nondefense applications.

The Air Domain

There are indications that the naval air segment is increasingly using COTS technologies and platforms. The Navy plans to use COTS platforms for medium UAS, and for large UAS, it plans to use COTS systems for the airframe with potentially specialized payloads.[48]

There are few extra-large UAS platforms dedicated to nondefense purposes. One of our interviewees speculated that since large platforms are subject to FAA regulations, only limited,

---

[48] Military UxS program official, interview with the authors, 2021. See Appendix B for descriptions of UxS size categories.

high-value commercial applications, such as the film industry, can justify the added expense. The AUVSI database lists 34 missions for UxS platforms and which ones each platform is marketed to perform, with each platform marketed to perform 8.6 missions on average. Figure 3.3 shows the number of platforms capable of performing each mission set, by type and size.

Figure 3.3. Number of U.S. Platforms for Each Mission, by Size and Use

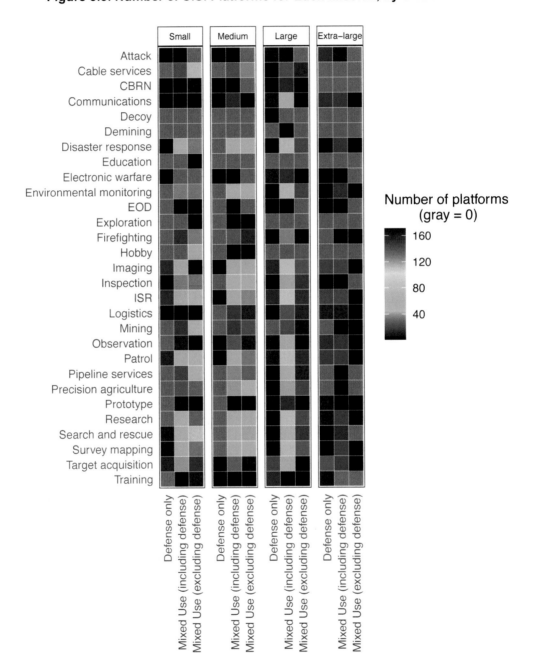

SOURCE: Authors' analysis of data from AUVSI.
NOTE: CBRN = chemical, biological, radiological, or nuclear; ISR = intelligence, surveillance, and reconnaissance.

Generally, as platform size increases, there are fewer serviced nondefense missions, as well as fewer platforms available per mission. For small UAS platforms, there are mixed-use platforms (excluding defense) for almost every mission, except for defense-oriented missions such as attack, decoy, demining, and electronic warfare. Furthermore, for many mission areas there are more small platforms available for nondefense markets than for defense markets. Medium platforms are more dual-use than small platforms, and there are more mission sets only serviced by dual-use platforms. Large platforms show a distinct transition toward defense markets. While there still are more dual-use large platforms than defense-only platforms for most missions, there are far more missions for which there are defense-only large platforms. At the same time, there is a decline in the number of nondefense missions services by large platforms. The defense orientation of extra-large platforms is especially notable. There are 14 missions for nondefense platforms marketed, but, as shown in Table 3.6, if nondefense platforms are further restricted to commercial-only platforms (excluding civil- and academic-use platforms), then there are only five mission sets serviced with any platforms: imaging, logistics, mining, prototype, and research.

Table 3.6. Number of Extra-Large UAS Platforms Marketed for Each Mission, by Commercial Market Type

Mission	Commercial (Excluding Defense)	Commercial (Including Defense)	Defense Only
Attack	0	2	19
Cable services	0	0	0
CBRN	0	0	0
Communications	0	4	32
Decoy	0	0	0
Demining	0	0	0
Disaster response	0	7	21
Education	0	0	0
Electronic warfare	0	0	6
Environmental monitoring	0	5	17
EOD	0	1	2
Exploration	0	0	0
Firefighting	0	2	5
Hobby	0	0	0
Imaging	1	8	44
Inspection	0	2	5
ISR	0	8	46
Logistics	3	12	22
Mining	1	2	1
Observation	0	3	33
Patrol	0	4	37
Pipeline services	0	3	3

Mission	Commercial (Excluding Defense)	Commercial (Including Defense)	Defense Only
Precision agriculture	0	3	3
Prototype	2	3	10
Research	1	9	27
Search and rescue	0	6	17
Survey mapping	0	5	21
Target acquisition	0	4	43
Training	0	0	6
Total	8	93	420

SOURCE: Authors' analysis of data from AUVSI, undated a.
NOTE: We excluded inactive platforms and any platforms with insufficient data.

The Maritime Domain

We found that the maritime domain had a mix of commercially available and defense-unique systems depending on the component. Maritime UxS generally use commercial propulsion systems, and there are multiple options available. Likewise, radar is generally not a defense-unique component, but a maritime system could require a radar to meet stricter specifications or have higher tolerance levels.

On the other hand, software and C2 radios are tailored to mission needs and are more likely to be considered defense unique. It is harder to find COTS systems for automation and payloads because some payloads are built to specifications set by the program office. Similar to the air domain, there is a strategy to use COTS systems where possible in the maritime domain.[49]

Findings

A mix of commercial demand and use implies a healthy market for UAS and UGSs, with the UUV market being smaller relative to those for UAS and UGSs. The USV market is smaller than the others, and while it may not be defense unique, it has few platforms overall.

It is important to define a commercial market platform in this context. In our view, a commercial market platform is one that has nondefense use cases. While the commercial market may appear healthy, commercial platforms are not necessarily readily transferrable (i.e., off the shelf) to defense use cases for regulatory reasons, because of acquisition limitations, or because there is no clear demand signal from DoD. We discuss how these issues manifest as defense-specific design requirements in the next section.

Defense Design Requirements

In this section we look at the degree to which defense design requirements might differ from commercial requirements for platforms that could otherwise serve both defense and nondefense

---

[49] Military UxS program official, interview with the authors, 2021.

44

markets. In some cases, certain defense-specific design requirements mean that only a subset of commercial platforms have the necessary design features to meet military requirements. In other cases, some defense-specific modification is required to make a platform that is otherwise commercially available suitable for military purposes.

The Air Domain

We found that the air domain had a mix of defense-unique design requirements. For example, on larger platforms, like the legacy RQ-4, there were no specialized design considerations for onboard battery cells, but there were for the packaging used to house the battery cells.[50]

Other components or software packages were found to require military-specific expertise. AI training algorithms in defense-specific environments is one example. Cybersecurity system design also involved military-specific requirements.[51]

In another case, we found that the operational environment influenced the design of systems rather than a specific military need. For example, the Navy's UAS typically use diesel or JP8 fuel due to the exclusion of gasoline onboard ships. Though it was entirely possible to source a gasoline-powered capability from commercial suppliers, specific requirements onboard ships artificially reduce the availability of suppliers due to the operational environment.[52]

The Maritime Domain

Within the maritime domain, we again found a mix of defense-specific design considerations. There were commercial designs incorporated—for example, USV engines were typically sourced from commercially available systems. However, the Navy has specific requirements that are more stringent and can rule out some commercial options. The case is similar for radar. There are commercially available radars, but the Navy requires higher reliability, among other needs. Cybersecurity requirements also create barriers to entry for commercial systems. INSs are yet another example.[53] Defense-specific knowledge is often needed at the system integration level, even if the components are commercially available—for example, for mine warfare systems. Even seemingly standard commercial items, such as paint, may need to meet specific requirements for defense use.[54]

Finally, part of the challenge in escaping the "valley of death" from initial proof of concept to deployed design is proving the capability is viable in an operational context, which may

---

[50] Military UxS program officials, interview with the authors, 2021.

[51] Industry representative, interview with the authors, 2021.

[52] Industry representatives, interview with the authors, 2021; military UxS program official, interview with the authors, 2021.

[53] Industry representatives, interviews with the authors, 2021.

[54] Industry representatives, interview with the authors, 2021.

require rethinking existing approaches to operations and sustainment.[55] For USVs and UUVs, that may include determining how to do maintenance with no humans on board, which will likely involve increased use of maintenance yards.[56]

The Ground Domain

In the ground domain, the operating environment, mission sets, and C2 requirements all affected the design considerations and physical capabilities demanded of the system. Some vendors reportedly did not comply with the Army's programming language, M-code, which restricted the pool of available vendors.[57]

Findings

Although we cannot prove it quantitatively, interviewees reported that defense design requirements are a significant problem area for the DIB. It is certainly easy for DoD and the services to say they want to leverage COTS systems, but the direct use of such systems in defense practice (or operations) seems very mixed—or use case dependent. As a result, there is a segment of the DIB that is effectively a defense-ready COTS system, where platforms are not truly off the shelf but are instead commercial platforms that are adapted in various ways or built purposefully such that they may meet defense-specific requirements before having a defense buyer. DIU's Blue UAS effort to continuously vet platforms to produce a policy-compliant set is a perfect example of a process to develop defense-ready COTS systems. The effort seems exclusively focused on small UAS and should be expanded to other platform types.

Quantifying the level of defense design specificity was a challenge beyond the scope of this effort, but it is something for OUSD(R&E) to consider in the future.

Skilled Labor Requirements

In this section we examine the degree to which specialized workforce skills are needed and available to integrate, manufacture, or maintain UxS capabilities. This includes touch labor as well as engineering and design labor. We relied on input from our interviews because we were unable to develop a quantitative approach.

All Domains

Although we discuss some domain-specific labor challenges, there are two labor challenges that are common across all UxS domains.

---

[55] The "valley of death" references "the period from inception to product market fit, during which a startup has to rely largely on money from investment to survive" and where products often fail; see Venture Collective, "The Valley of Death," undated.

[56] Justin Katz, "For Unmanned Vessels, Navy Still Working Out Maintenance Strategy," *Breaking Defense*, January 12, 2022a.

[57] Military UxS program official, interview with the authors, 2021.

The first challenge is attracting and retaining personnel in STEM fields, particularly in the areas of AI and cybersecurity. Three interviewees raised this issue. To a large extent, the UxS DIB faces the same challenges as many other industries competing for STEM talent: there often are more STEM jobs than STEM professionals, regardless of subfield or industry. Competition is tight, since firms in the UxS DIB compete with large technology companies for the same talent, such as software engineers.[58] Interviewees identified two STEM subfields that are more specialized and therefore might have a smaller pool of competing employers, but also have a smaller pool of qualified professionals: AI and cybersecurity.

The challenge of attracting and retaining early-career and midcareer STEM professionals is especially costly in the DIB, since there is some indication that defense-specific experience is helpful for some defense UxS design stages, such as integrating sensors as payloads. When STEM professionals leave for commercial technology firms, it is difficult to replace them with similarly experienced professionals. The reliance of some DoD legacy systems on components that are obsolete in the commercial sector exacerbates some labor challenges. For instance, representatives from one program office that we spoke with stated that for some replacement parts, there is only one person at one company who knows how to make an obsolete part, which is purchased in small batches. Several industry experts also suggested that obsolescence increases the challenge of retaining STEM workers, such as specialists in cutting-edge technologies, who tend not to stay in jobs that do not leverage and expand their knowledge of those technologies.[59]

Another labor challenge across domains pertains to workers with industry-specific knowledge. Representatives from two firms, representing two different UxS domains, stated that they struggled to find enough workers with training and experience in certain industries underlying UxS platforms, such as high-end machining and electronics.[60]

The Maritime Domain

USVs and UUVs have some unique labor challenges. Representatives from multiple program offices and vendors identified shortages of welders and forging skills as challenging to USVs and UUVs, especially for larger platforms. Shipbuilding for USVs and UUVs is perhaps the most challenging labor environment as regards retaining welders.[61]

Findings

We were unable to draw comprehensive quantitative findings or independently verify vendor assertions, but vendors and government program offices alike identified issues with finding and

---

[58] Industry representatives, interviews with the authors, 2021; military UxS program officials, interviews with the authors, 2021.

[59] Industry representatives, interviews with the authors, 2021; military UxS program officials, interviews with the authors, 2021.

[60] Industry representatives, interviews with the authors, 2021.

[61] Military UxS program officials, interviews with the authors, 2021.

retaining STEM professionals, such as AI and cybersecurity specialists, as well as skilled labor with UxS experience, such as for high-end machining and welding. The UxS DIB is not immune to labor challenges affecting the broader economy, and uncertainty in procurement levels does not help vendors attract and retain talent, even though many of the technologies are on the cutting edge and potentially attractive to workers.

Facilities and Equipment

In this section we look at whether specialized equipment or facilities are needed to integrate, manufacture, or maintain UxS capabilities.

The Air Domain

The air domain demonstrated a few challenges related to testing facilities. Navy programs for large and long-range platforms had difficulties scheduling their evaluations. Specifically, there were challenges with respect to getting access to radio frequency ranges to test the required spectrums in different environments. Currently there are no other options for this type of testing, making it difficult to schedule time at the test facility.[62] No problems were reported for test ranges tailored to smaller UAS platforms.

The Maritime Domain

In the maritime domain, stakeholders reported greater challenges when trying to access testing facilities. One reason was that some testing facilities were located at joint bases where one service had the authority to determine priority of access to radio frequencies and testing areas. For example, during one test, the frequency required for a maritime unit's control systems and safety operator kill switches conflicted with one in use by the Air Force; because the Air Force was the authority on the base, the Navy's systems received less priority and access. This led one interviewee to state that the Navy's testing infrastructure was too limited. Others mentioned testing restrictions on autonomous systems at full ranges and operationally relevant durations in the ocean for USVs and UUVs. Finally, ranges for larger USV and UUV platforms testing long-range, long-endurance mission sets were characterized as more difficult to manage.[63]

The testing needed for UUVs also requires unique equipment like hyperbaric chambers and acoustic test chambers in addition to boating facilities to house the platforms.[64]

Findings

Our interviewees suggested that there is limited capacity at UAS ranges that allow the broad use of radio frequency testing, and particularly those capable of simulating electronic warfare

---

[62] Military UxS program officials, interviews with the authors, 2021.

[63] Industry representatives, interview with the authors, 2021; military UxS program officials, interview with the authors, 2021.

[64] Industry representatives, interviews with the authors, 2021.

conditions. Additionally, scheduling for large UAS platforms could be an issue, suggesting capacity DIB challenges.

There are indications that the Navy's USV and UUV testing infrastructure is also limited in capacity. Maritime platforms have encountered challenges in finding opportunities to complete full-range autonomy testing at operationally relevant durations. Furthermore, many prototype long-range USVs today are retrofitted platforms designed for maintenance at sea; because the platforms are now uncrewed, additional facilities will likely be needed to provide maintenance.

Long-Lead-Time Materials and Components

In this section we examine whether there are materials, components, or other required items that take substantially longer to manufacture or obtain than other system inputs. These items with long lead times represent constraints in the supply chain that can slow down deliveries.

All Domains

The COVID-19 pandemic has led to delays for certain types of products used in uncrewed platforms in all domains. Three such issues are as follows:

1. Pandemic-driven shortages of microchips, electronics, and computer components were mentioned in four interviews.[65] U.S. Department of Commerce data showed microchips "typically had a lead time—the time from order to delivery of the product—of between 84 and 182 days. By late 2021, that had doubled for some key products, and the lead time had stretched to 103–365 days."[66] Smaller UxS manufacturers were unable to compete with larger bulk buyers, indicating that semiconductor manufacturers naturally gravitated to the bigger purchases. Some manufacturers depended on (or had had to cultivate) relationships with second-tier semiconductor suppliers to acquire materials. Companies also found sensors hard to source within reasonable timelines because of the microchip shortage.[67]
2. Delays in deliveries of specialized batteries also posed a challenge. All of our interviewees asserted that they had suffered pandemic-related delays, with lithium-ion batteries among the commonly mentioned components.[68] Batteries can have long lead times because they require specialized packaging, and small fleet sizes provide little incentive for suppliers to prioritize orders, particularly in light of demand from the electric vehicle market. According to our interviewees, legacy platforms are particularly susceptible to these concerns because orders of spares or replacements are small. Vendors often wait until

---

[65] Industry representatives, interviews with the authors, 2021; DIU officials, interviews with the authors, 2021; military UxS program officials, interview with the authors, 2021.

[66] Josh Zumbrun and Alex Leary, "Chip Shortage Leaves U.S. Companies Dangerously Low on Semiconductors, Report Says," *Wall Street Journal*, January 25, 2022.

[67] Industry representative, interviews with the authors, 2021; military UxS program officials, interview with the authors, 2021.

[68] Industry representatives, interviews with the authors, 2021; DIU officials, interviews with the authors, 2021; military UxS program officials, interviews with the authors, 2021.

they have enough orders to batch-produce components in larger quantities to justify the cost of tooling, labor, and material sourcing.[69] Lithium-ion batteries also contain five materials on the U.S. Geological Survey's critical minerals list.[70]

3. Delivery delays in materials, such as fiberglass and ferrite, were common because of limited suppliers and other difficulties procuring materials and fabricating components.[71]

The Air Domain

We found materials with similar constraints in the air domain. Lightweight carbon composite materials are used for airframes, and interviewees reported limited sources for larger frames. These composites are typically woven and layered into sheets in a method that results in long lead times.[72]

The Maritime Domain

In the maritime domain, unique capabilities rely on several critical materials. For instance, undersea cables and connectors (used by USVs and UUVs) require iridium, a transition metal listed as a critical mineral.[73]

Because UUVs are subjected to high pressure at low depth, they require strong hull materials like titanium. Though they rely on common technology for production, the titanium these larger platforms rely on is more difficult to source and work with. More common metals like aluminum and welding wire have also experienced price increases, and increased delays, due to the pandemic.[74]

Findings

Lead times affect the capacity of the UxS DIB in several ways. Specifically, the DIB is not immune to ongoing supply chain issues with microchips, lithium-ion batteries, and raw materials such as fiberglass and ferrite.

There may be an opportunity for DoD to leverage better buying-power initiatives to support vendors in competition with other industries.

Availability of Alternatives

In this section we review whether cost-, time-, and performance-effective alternatives meet DoD needs.

---

[69] Military UxS program officials, interview with the authors, 2021.

[70] They are cobalt, graphite, lithium, manganese, and nickel. See Jason Burton, "U.S. Geological Survey Releases 2022 List of Critical Minerals," press release, Reston Va.: U.S. Geological Survey, February 22, 2022.

[71] Military UxS program officials, interview with the authors, 2021.

[72] Industry representatives, interviews with the authors, 2021.

[73] Military UxS program officials, interviews with the authors, 2021; Burton, 2022.

[74] Industry representatives, interviews with the authors, 2021.

All Domains

Turning again to the AUVSI data, there are in general many different UAS platforms capable of performing the same types of missions, as Table 3.7 shows. Although few UGS platforms are capable of surveying and mapping, several platforms are available for other types of missions. There seem to be several USV-relevant mission areas with relatively few platforms: CBRN (three platforms); electronic warfare (one platform), and logistics (five platforms). Most UUV mission areas seem to be relatively well served by alternative platform options.

Table 3.7. Count of U.S. Platforms Capable of Performing Selected Missions

Mission	UAS	UGS	USV	UUV
Attack	37	15	7	0
Cable services	193	5	0	36
CBRN	31	37	3	7
Cleaning	0	11	0	10
Communications	119	18	7	4
Construction	0	1	1	28
Decoy	3	0	0	1
Demining	4	7	7	25
Disaster response	406	67	5	13
Education	3	10	0	2
Electronic warfare	36	0	1	0
Environmental monitoring	384	6	32	104
EOD	11	49	6	29
Exploration	5	85	2	47
Firefighting	143	3	1	0
Health care	0	8	0	0
Hobby	62	5	0	2
Household domestic	0	5	0	0
Humanoid	0	2	0	0
Imaging	550	18	14	32
Inspection	331	112	13	139
ISR	415	105	32	43
Logistics	100	126	5	2
Maintenance	0	6	0	46
Mining	159	10	7	26
Observation	156	23	10	42
Obstacle clearance	0	16	0	0
Patrol	370	97	32	47
Pipeline services	200	9	7	61
Precision agriculture	316	7	0	0
Prototype	51	6	2	5
Research	431	39	34	107
Salvage	0	0	3	75

Mission	UAS	UGS	USV	UUV
Search and rescue	371	61	10	20
Survey mapping	418	8	30	103
Target acquisition	226	68	9	4
Training	42	2	2	2

SOURCE: Authors' analysis of data from AUVSI, undated a.
NOTES: We excluded inactive platforms and any platforms with insufficient data.

Figure 3.4 further explores available alternative platforms for missions by examining the number of platforms of each size for each mission set. Given the limited information about the precise capabilities of the platforms, size might provide more information about their mission capabilities. For example, a large UGS might be capable of moving larger and heavier equipment than a small UGS. Figure 3.4 shows the number of platforms on a color scale indicating the rough number of platforms for each mission, size, and type.

Figure 3.4. Number of U.S. Platforms for Each Mission, by Type and Size

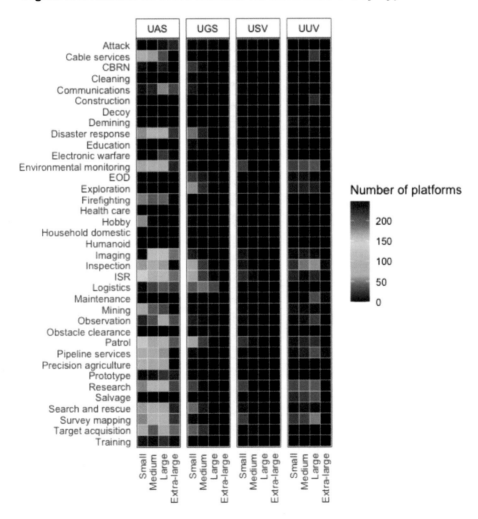

SOURCE: Authors' analysis of data from AUVSI.

52

UAS missions generally have more platforms available, including multiple platforms in each size category. However, there are relatively few small UAS attack, communications, and logistics platforms, and relatively few large UAS electronic warfare platforms. Across the remaining domains, the availability of alternative platforms generally decreases as platform size increases.

At the component level, interviewees suggested that there were issues with the availability of alternatives for batteries not sourced from China, composites, microchips, motherboards, radar, and carbon fiber. We were unable to find corroborating evidence to confirm or deny these assertions.

The Maritime Domain

One first-tier supplier indicated that INSs depended on a single domestic supplier because of the higher accuracy needed for maritime systems compared with aerial applications.[75] Additionally, sources of titanium seemed limited, and interviewees identified challenges with large UUV forging capacity.

Findings

In terms of first-tier suppliers, we see that generally, there are many different UAS platforms of different sizes capable of performing the same types of defense-related missions. Although few UGS platforms are capable of surveying and mapping, several platforms are available for other types of missions. There seem to be several USV-relevant mission areas with relatively few platforms: CBRN, electronic warfare, and logistics. Most UUV mission areas seem to be relatively well served by alternative platform options. Across the ground and maritime domains, the availability of alternative platforms generally decreases as platform size increases.

We observed issues predominantly with second- and third-tier suppliers and supplies when it came to the availability of alternatives. Interviewees asserted that there were problems finding sources of second-tier items, such as batteries from countries other than China, microchips, radars, and some payloads, such as maritime INSs. Third-tier availability issues were noted for non-Chinese lithium, carbon fiber, and titanium (for UUVs). We also were also told that large UUV forging capacity was limited. However, we were unable to corroborate these assertions with other evidence.

---

[75] Industry representative, interview with the authors, 2021.

Chapter 4. Assessment of Chinese Uncrewed Defense Industrial Base Systems

China has increased its focus on UxS in the past two decades and is eager to learn from and improve on the work of other countries to advance to the forefront of UxS technology. Internationally, China frequently touts the technological strides it has made, and it is clear that UxS are becoming a key area of emphasis.[1] Domestically, Chinese academics have been arguing for an increase in UxS use as a measure to minimalize casualties on the battlefield.[2] China classifies UxS according to three "combat systems": the ground unmanned combat system, the air unmanned combat system, and the sea unmanned combat system. These systems can be divided into four categories of autonomy: remote control, semiautonomous, platform-centric autonomous, and network-centric autonomous.[3] In this chapter we discuss China's current and future position in manufacturing and fielding each type of UxS, and any known deficiencies that China is facing.

Demand Signal: Expected Units and Production Cost

We used Janes Markets Forecast data to examine China's expected demand signal for the near future, as shown in Figure 4.1. These data should be taken with a grain of salt due to the lack of transparency associated with data on expected Chinese military procurements.

China is planning to increase the number of military communication upgrades to its legacy platforms, as well as new UAS, but the increases in UGS, USV, and UUV quantities are limited. Figure 4.2 shows the increases in USV and UUV platforms on a smaller scale.

The expected increases in production costs for each type of UxS are shown in Figure 4.3.

---

[1] Michael S. Chase, Kristen A. Gunness, Lyle J. Morris, Samuel K. Berkowitz, and Benjamin S. Purser III, *Emerging Trends in China's Development of Unmanned Systems*, Santa Monica, Calif.: RAND Corporation, RR-990-OSD, 2015.

[2] As Wang Xingcheng and Chen Hai, "Application of Unmanned Combat System and Research on Key Issues," *Military Digest*, Vol. 7, April 2021, p. 23 argue,

> Moreover, with the development of society and the stability of people's lives, the country's ability to withstand casualties in war has been steadily weakened. Therefore, the development of unmanned combat equipment instead of manned combat to avoid direct damage to combat personnel by lethal weapons and reduce casualties has become an urgent task facing military powers.

[3] Wang and Chen, 2021.

Figure 4.1. Chinese Historical and Expected Military UxS Units Delivered, 2016–2026

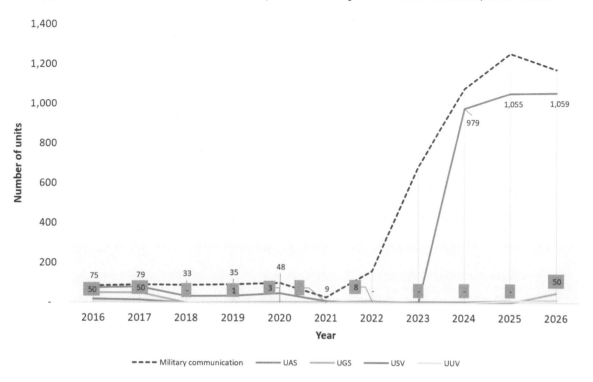

SOURCE: Authors' analysis of data from Janes Markets Forecast.
NOTES: The Military Communication category consists primarily of upgrades to legacy systems. UGS platform counts have an orange background.

Figure 4.2. Chinese Historical and Expected Military UxS Units Delivered, 2016–2026: USVs and UUVs

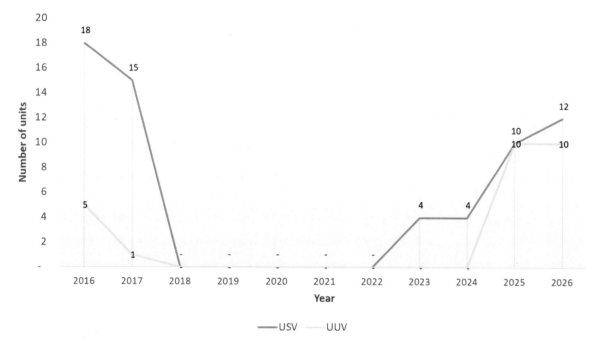

SOURCE: Authors' analysis of data from Janes Markets Forecast.
NOTE: The scale has changed from the previous figure.

Figure 4.3. Chinese Military UxS Expected Production Costs, by Janes Type and Use (in millions of dollars)

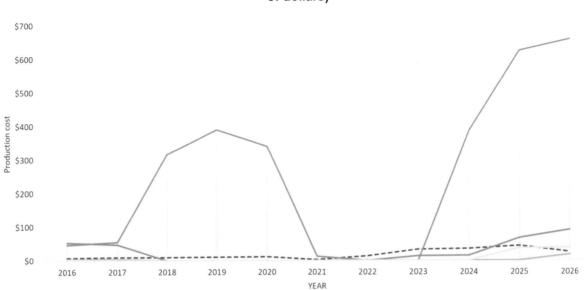

SOURCE: Authors' analysis of data from Janes Markets Forecast.
NOTE: The Military Communication category consists primarily of upgrades to legacy systems.

Total production costs for UAS were expected to increase from $16.58 million in 2021 to $843 million in 2026, though 2021 may be an aberration given that the average from 2016 to 2020 was $263 million; it is likely that there was additional investment not captured in the Janes data, and the data also indicate significant expected increases in investment in UAS relative to other types of UxS.

Uncrewed Aerial Systems

China is interested in all areas of UxS technology, but it has made more progress in some domains than others, with China's UAS program (active since the mid-1990s) taking a significant lead.[4] By 2011, the People's Liberation Army (PLA) had fielded a total of 208 UAS.[5] More recently, China has moved toward developing UAS as a result of economic incentives and these systems' contributions to China's military strength.[6] China is estimated to hold more than 70 percent of the global market share of commercial UAS, with its largest commercial drone

---

[4] Defense Science Board, *The Role of Autonomy in DoD Systems*, Washington, D.C.: U.S. Department of Defense, July 2012.

[5] Ian M. Easton and L. C. Russell Hsiao, *The Chinese People's Liberation Army's UAV Project*, Arlington, Va.: Project 2049 Institute, OP 13-001, 2013.

[6] Furthermore, this trend is expected to continue; see DoD, *Counter–Small Unmanned Aircraft Systems Strategy*, Washington, D.C., 2020a.

manufacturer, DJI, accounting for the vast majority of this share.[7] As China's UAS technology and ability to domestically produce these platforms continues to grow, the PLA Air Force appears to be making strides catching up to its Western counterparts.[8] China has been conducting more exercises with UAS, including a 2019 bilateral exercise with Tajikistan and 2021 counterterrorism military drills with India, Kazakhstan, Pakistan, Russia, Tajikistan, and Uzbekistan.[9]

China's achievements in autonomous UAS are among the most large and complex in the world; academics predict that these improvements will be used to bolster China's territorial claims in contested regions such as the South China Sea.[10]

Due to China's mature UAS industry, along with the low cost of developing small drones, China's UAS are the most advanced of China's UxS.[11] The world's largest commercial drone manufacturer is DJI, based in Shenzhen, a city with more than 600 commercial UAS companies.[12] Furthermore, U.S. sources note that China is one of a small number of countries developing high-speed combat UAS.[13] China is also developing drone swarming technology, for which it has filed a large share of recent patents.[14] Researchers have noted that UAS are advantageous because they are lightweight, small, low-cost, and highly maneuverable, and they can be operated safety; can reliably take off autonomously; and can be programmed for automatic follow, pointed flight, automatic return, and automatic landing.[15]

Issues

However, several Chinese articles we reviewed mentioned concerns about China's UAS industry. Shulin Yang, Xiaobing Yang, and Jianyou Mo note that there are problems with UAS

---

[7] As a result, China claims to be the market leader in commercial drones, which, it notes, is key to setting standards in the drone industry and "[defining] the future UAS landscape." See Cheng Yu, "China Now 'Leader' in Unmanned Aircraft System," *China Daily*, September 1, 2020.

[8] OSD, *Military and Security Developments Involving the People's Republic of China*, Washington, D.C., 2020.

[9] The Chinese Ministry of National Defense has stated that the counterterrorism drill drew on combat experience from the Syrian War and the Armenia-Azerbaijan conflict over the disputed Nagorno-Karabakh region. See Minnie Chan, "Drone Warfare Marks a First for SCO Drill, as Region Faces Up to Terror Attack Risks in Afghan Fallout," *South China Morning Post*, October 3, 2021; and OSD, 2020.

[10] OSD, 2020; Easton and Hsiao, 2013.

[11] Chad J. R. Ohlandt and Jon Schmid, "Flying High: Chinese Innovation in Unmanned Aerial Vehicles," paper presented at the CAPS-RAND-NDU Conference on the People's Liberation Army, Arlington, Va., April 30, 2018.

[12] Masha Borak and Yujie Xue, "How Shenzhen, the Hi-Tech Hub of China, Became the Drone Capital of the World," *South China Morning Post*, April 4, 2021.

[13] U.S. Government Accountability Office, *Nonproliferation: Agencies Could Improve Information Sharing and End-Use Monitoring on Unmanned Aerial Vehicle Exports*, Washington, D.C., GAO-12-536, July 30, 2012.

[14] Furthermore, recent patents for swarming technologies are heavily skewed toward Chinese companies, dominated by Beihang University; see Ohlandt and Schmid, 2018.

[15] Sun Yongsheng, Jin Wei, and Tang Yuchao, "Application of Unmanned System in Prevention and Control of COVID-19," *Science and Technology Review* (China), Vol. 38, No. 4, 2020.

development in China stemming from a lack of administrative policies; regulations; research on supporting technologies; trained professionals; and industry standardization, strategies, and protocol.[16] In addition, the areas of counter-UAS capabilities, which may include UAS, are also a point of discussion. A researcher from Yuan Wang Military Science and Technology Institute in Beijing argued that while drones are becoming more combat ready, preventing and detecting UAS attacks is "still a new subject to China."[17] Yang Wei and colleagues note that Chinese air defense is lagging and that China needs to "achieve leap-forward development" of uncrewed and autonomous air defense systems.[18] Although the PLA is focusing on the swarm intelligence and teaming capabilities of UAS, independent observers argue that China can enhance communication, ensure resilient and self-healing swarms, and improve swarm resilience to stay in the air longer.[19]

Uncrewed Ground Systems

There is very little discussion in Chinese literature about developments in the ground domain. Wang Xingcheng and Chen Hai note that the complexity of the ground environment is significantly higher than the air and sea domains and thus development "lags far behind" the air and sea.[20]

Uncrewed Surface Vehicles and Uncrewed Underwater Vehicles

Chen notes that the missions of USVs include continuous patrolling, high-quality images, reduction in casualties, and immediate deployment "in enormous numbers to overwhelm the enemy."[21] Lin Long-xin and Zhang Bi-sheng claim that USVs are in a stage of "rapid development" and "will have a profound and even revolutionary impact on future maritime operations." However, the systems are still based on preprogrammed control and remote control models, and "there are still major deficiencies in the ability to effectively respond to uncertain

---

[16] The authors note that variations in UAS types and sizes complicate production, inhibit regulations that allow for government subsidies, and hinder compatibility, which prevents repairs and raises costs; see Shulin Yang, Xiaobing Yang, and Jianyou Mo, "The Application of Unmanned Aircraft Systems to Plant Protection in China," *Precision Agriculture*, Vol. 19, No. 2, April 2018.

[17] Chan, 2021.

[18] Yang Wei, Wang Yue, Liu Xuechao, Zhao Kai, Xue Peng, and Zhang Bo, "Development and Research Summary of Unmanned Air Defense Weapons," *Journal of Gun Launch and Control*, Vol. 4, March 6, 2021.

[19] Tate Nurkin, Kelly Bedard, James Clad, Cameron Scott, and Jon Grevatt, *China's Advanced Weapons Systems*, London: Jane's.by IHS Markit, May 12, 2018; OSD, 2020.

[20] Wang and Chen, 2021.

[21] Stephen Chen, "After Drones, China Turns to Unmanned Vessels to Boost Its Marine Power," *South China Morning Post*, December 5, 2013. The article also notes that USVs should be able to be deployed in large numbers across Chinese waters near neighboring countries.

and complex environments and to deal with emergencies."[22] China has announced that several 100-foot-long extra-large UUVs are in development, with a predicted deployment within the decade.[23] China's interest in UUVs has increased dramatically in recent years, with most of the funding coming from the PLA, and it is focused on imaging systems and communication technology.[24]

Issues

Stephen Chen notes that surveillance, a major aspect of USV missions, is difficult for Chinese USVs due to technical issues with developing a camera system that can take sharp and stable pictures, both in the hardware to stabilize the camera and software to compensate for blur.[25] Beyond surveillance issues, Lin and Zhang note that due to the unpredictability of the sea surface and limited technological development, Chinese USV technology is still oriented toward researching combat applications and testing.[26] Other concerns include deployment and recovery, as well as "excessive dependence" on satellite navigation, unstable communication, and inability to respond to changing conditions.[27]

Zhang Xishan, Lian Guangyo, and colleagues note that, across all domains, development of autonomous uncrewed support equipment is at a low level and research is fragmented, with a lack of top-level developmental planning, creating a gap between Chinese development and other nation's capabilities.[28]

Interest in International Uncrewed Systems

Chinese academics have expressed interest in the advances other countries have made in various UxS domains. These articles offer interesting insights into the areas of the international

---

[22] Lin Long-xin and Zhang Bi-sheng, "Technical Development and Operational Application of an Unmanned Surface Combat System," *Journal of Unmanned Systems*, Vol. 26, No. 2, 2018.

[23] The Chinese XL-UUVs, HSU-001s, have only been displayed during military parades. As a result, the actual readiness of this technology is unclear. Due to the complexity of deployment in the underwater domain, observers express reservations regarding HSU-001 operational capability. See J. R. Wilson, Unmanned Submarines Seen as Key to Dominating the World's Oceans," *Military and Aerospace Electronics*, October 15, 2019; and David Hambling, "China's New Unmanned Attack Sub May Not Be What It Seems (Update: In Fact It's A Paper Tiger)," *Forbes*, July 9, 2021b.

[24] Chase et al., 2015.

[25] Chen, 2013.

[26] Lin and Zhang, 2018.

[27] Wang Xingcheng and Chen Hai, "Application of Unmanned Combat System and Research on Key Issues," *Military Digest*, Vol. 4, April 2020.

[28] Zhang Xishan, Lian Guangyao, Li Huijie, and Yuan Xiangbo, "Development and Application of Intelligent Unmanned Support Equipment," *National Defense Technology*, Vol. 41, No. 2, April 2020.

UxS industry where China focuses the most interest. Chinese authors list 20 U.S. UAS by name, including the Scan Eagle and Global Hawk; five UGS platforms; four UUVs; and three USV platforms. Additionally, Chinese authors list six Russian UxS; four British UxS; four Israeli UxS; and one system each from France, Germany, Italy, and Turkey. Furthermore, Chinese analysts have been paying considerable attention to U.S. UxS deployment in Afghanistan, Iraq, Libya, and Pakistan and frequently mention the advantage that arises from "battle-tested" UxS.[29]

More generally, there is significant rhetoric on international advancement in the UxS domains. Liu Haijiang, Li Xiangang, and Liang Ming estimate that, by 2025, UxS will account for more than 30 percent of the Russian military's equipment structure and that by 2040 more than half of the U.S. military equipment on the battlefield will be UxS. They also state that U.S. UxS are moving toward clustering, actual combat, and both autonomous and intelligent decisionmaking.[30] Zhang, Lian, and colleagues note that the United States leads the world in technical research, software and hardware equipment, and practical applications for intelligent uncrewed support equipment.[31] The Defense Science Board's 2012 analysis of China's UxS production noted that although China openly looks to the West for technical expertise and suppliers in the commercial sector, the same cannot be done for the defense industry—and this has led China to copy successful U.S. designs to accelerate its UxS development.[32] However, Tate Nurkin and colleagues believe that China is no longer relying on copying U.S. designs, despite China's statements that its indigenous innovation capability is an issue for its UxS industry.[33]

Chinese academics offer many suggestions in domestic reports on ways the PLA can improve its UxS capabilities in order to compete with these international examples. Liu, Li, and Liang argue that the PLA must unite with universities, research institutes, and military-industrial

---

[29] Jonathan Ray, Katie Atha, Edward Francis, Caleb Dependahl, James Mulvenon, Daniel Alderman, and Leigh Ann Ragland-Luce, *China's Industrial and Military Robotics Development*, Washington, D.C.: U.S.-China Economic and Security Review Commission, October 2016; Wang and Chen, 2021.

[30] Liu Haijiang, Li Xiangang, and Liang Ming, "A Thinking on Accelerating the Development of Unmanned Weapon Systems and Technologies," *National Defense Technology*, Vol. 41, No. 6, December 2020.

[31] Zhang, Lian, Li, and Yuan, 2020. The report mentions that the U.S. military currently fields 7,500 UAVs and 15,000 UGVs, which form a "comprehensive" system of unmanned equipment (p. 1).

[32] Defense Science Board, 2012.

[33] Nurkin, Bedard, Clad, Scott, and Grevatt, 2018; Andrea Gilli and Mauro Gilli, "Why China Has Not Caught Up Yet: Military-Technological Superiority and the Limits of Imitation, Reverse Engineering, and Cyber Espionage," *International Security*, Vol. 43, No. 3, 2019, argue that it has become more difficult to copy advanced military technology, using China's struggle to imitate the F-22 Raptor as an example. Although it may be relatively easy to copy less sophisticated technologies, they assert that the most advanced technologies require tacit and organizational know-how based on extensive and expensive experimentation and testing and diverse expertise. As a result, China has not been very successful at copying advanced weapon systems due to the complexity of the technology and components, quality issues, and the high costs of cyberespionage and the implementation and production of copied designs.

enterprises to set up uncrewed combat disciplines and professional groups, as well as improve the combatants' skills.[34] Wang Wen-feng, Yu Xue-mei, and Xu Dong-mei have stated that the PLA must improve UxS interoperability by improving standards and developing UxS technology that can collaborate between products and systems to complete missions.[35] Through strong political and financial support, Chinese universities and companies across both the civil and military spheres are building UxS research centers to reach the level of foreign military UxS capabilities.[36]

Exports

China has fewer export restrictions for UAS than other countries, making it an attractive source from which developing countries can purchase UAS.[37] Since 2015, China's state-owned Aviation Industry Corporation (AVIC) has exported 50 Wing Loong II drones alone; these drones are considered "battle tested" after being used in Libya and Nigeria.[38]

To attract buyers, Chinese manufacturers emphasize quantity and lower prices over technical capabilities.[39] In 2019, AVIC was the sixth-highest-grossing seller of military equipment of all types, with $22.5 billion in sales, the highest ranked non-U.S. company.[40] Across both commercial and military UAV sales, OSD indicated that China is the second-largest exporter, and until 2017, China was even selling drones to the U.S. Army.[41] However, a report by a Chinese research institute noted difficulties faced by non-state-owned companies in entering the export market due

---

[34] Liu, Li, and Liang, 2020. The authors specifically mention that these initiatives should follow the same method as the Defense Acquisition Improvement Act of 1990, which highlights "awarding grants to public colleges or universities for the improvement of undergraduate or graduate education in scientific disciplines critical to the national security functions of DOD" as well as other improvements to research in national-defense-related areas; see Defense Acquisition Improvement Act of 1990, Congress, 101st, Congress, S.2916, 1990.

[35] Wang Wen-feng, Yu Xue-mei, and Xu Dong-mei, "Overview of Unmanned Systems Interoperability Standardization," *China Electronic Standardization Institute*, Vol. 12, No. 1, 2020.

[36] Ray, Atha, Francis, Dependahl, Mulvenon, Alderman, and Ragland-Luce, 2016.

[37] U.S. Government Accountability Office, 2012.

[38] The conflicts in which Wing Loong II drones have been involved include Nigerian battles with Boko Haram and the Libyan Civil War. See Bruce Einhorn, "Combat Drones Made in China Are Coming to a Conflict Near You," *Bloomberg Businessweek*, March 18, 2021.

[39] Ray, Atha, Francis, Dependahl, Mulvenon, Alderman, and Ragland-Luce, 2016.

[40] Einhorn, 2021.

[41] OSD, 2020. In August 2017, an Army memorandum ordered all units to remove DJI Chinese drone technology from use and replace it with U.S.-built equipment. However, U.S. companies noted that DJI's manufacturing capability made it difficult to compete. One representative noted, "It's just going to be inherently much more difficult for a Silicon Valley–based, software-focused company to compete against a vertically integrated powerhouse manufacturing company in China" (Mac Ryan, quoted in Nurkin, Bedard, Clad, Scott, and Grevatt, 2018, p. 157).

to difficulties in developing military trade, low investment, low contact with foreign countries, and inadequate support measures.[42]

In recent years, several news outlets and other commentators have reported that China is the leading exporter of combat UAS, based on data published by the Stockholm International Peace Research Institute (SIPRI). For example, Bruce Einhorn reports that between 2010 and 2020, China delivered 220 combat UAS to other nations, followed by the United States, which exported ten combat UAS.[43] In Figure 4.4 (reproduced from Einhorn), Chinese sales are shown in red and U.S. sales are in blue. However, we found that these reports were misleading. In particular, note that these numbers omit exports of ISR UAS.

Figure 4.4. Number of Combat UAS Delivered to Other Nations by China and the United States, 2010–2020

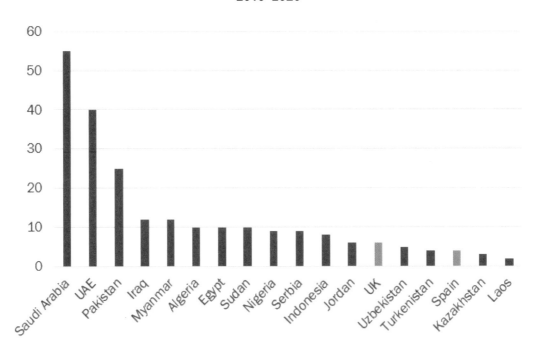

SOURCE: Einhorn, 2021.

We examined the SIPRI data and found that it indicates the United States exported 291 ISR UAS (including Scan Eagles, Blackjacks, Shadows, and Global Hawks) and 36 combat UAS (Reapers and Predators) from 2010 to 2020, while China exported 220 combat UAS (CH-3s,

---

[42] The company in question, ZYD Research Institute, issued the report to determine ways of increasing its exports. The author notes that ZYD's main export products are photoelectric products, including laser rangefinders, photoelectric fire control trackers, and photoelectric seekers. Measurement and control, radar, and drone products have not been exported. Shan Xiuxu, *Research on Trade in Military Products of ZYD Institute*, Zhengzhou, China: Zhengzhou University, 2016.

[43] Einhorn, 2021.

CH-4s, and Wing Loongs) and 18 ISR UAS (ASN-209s). Both types are considered "arms transfers" by SIPRI because they were intergovernmental sales for military purposes.[44]

Fragility and Criticality

China's UxS market is opaque, with all of China's major UxS companies linked to the Chinese government. We attempted to trace affiliations between UxS companies, universities, and Chinese government entities involved in our China platform sample set (see Appendix B); the results are shown in Figure 4.5. We found relationships between 61 organizations involved in the development of the 18 sample-set platforms. Forty-eight percent of these affiliated organizations are companies (shown in purple), 22 percent are universities (shown in green), 21 percent are government entities (shown in orange), and 9 percent are designated laboratories (shown in blue). These affiliations were collected by tracking parent companies through Bloomberg Business, searching company data stored on Janes, gathering news articles regarding mergers and partnerships, reading Chinese company registration documents available from CNKI, and

Figure 4.5. Chinese Company Affiliations of Sample-Set Manufacturers

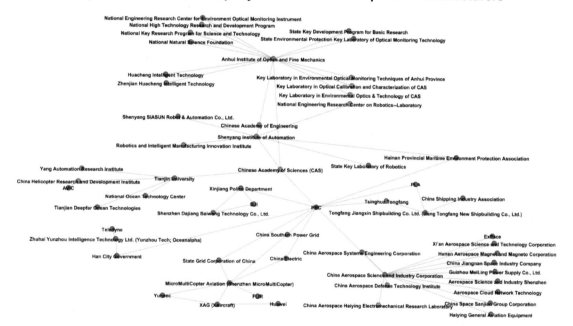

SOURCE: Authors' analysis of Chinese websites.

---

[44] There were also discrepancies between the SIPRI export data and the Janes military export data we used for the analysis reported in Chapter 6. Over a slightly longer period from 2009 to 2020, the Janes data indicated that U.S. export sales of defense UAS were $4.6 billion, relative to China's $195 million. Janes reported a total of 2,311 UAS exported by the U.S., including 291 ScanEagles, 55 Blackjacks, 52 Shadows, eight Global Hawks, and 42 Predators and Reapers; and 496 UAS exported by China, but only 79 CH-3s, CH-4s, and Wing Loongs. The remainder were small UAS exported to foreign militaries by DJI. See Appendix A, Table A.1. It was beyond the scope of this project to attempt to reconcile the differences.

visiting the company websites. Only affiliations with explicitly stated partnerships or financial support were recorded.

The CAS, displayed in the center of the graph, is a think tank, academic governing body, and source of academic funding and research. Its parent agency is the State Council of China, and it has been rated as the top institution of scientific research by the Nature Index since 2016.[45] Furthermore, Nurkin and colleagues note that Chinese state-sponsored enterprises and research institutes are assisted by the private sector in the development of military-purpose systems.[46] State-backed investment, combined with the fact that many Chinese company executives are Communist Party members, has blurred the lines between public and private industries in China.[47] Therefore, with its centralized, state-controlled research capabilities, China can exert a large amount of control over the trajectory of UxS R&D.

Beyond these affiliations and the clear overlap among private, military, and consumer industries, information on China's UxS industry is scant.[48] Zhang, Lian, and colleagues noted that some key components of UxS manufacturing still rely on imports, but the article did not elaborate.[49] A 2015 article noted that the People's Liberation Army Air Force moved target drone engine manufacturing to the private sector to avoid the bloated prices of state-owned companies, yet some components are beyond private-sector capability, including turbine blades.[50] However, we had little success tracking the sources of individual parts in the fused military-civil industry of China, in part as a result of the complexity of the systems. The Chinese Wing Loong II UAV alone is described as containing 130 million parts, with 60 percent of its parts provided by 300 private companies.[51]

An analysis of FactSet data on Chinese entities revealed some international suppliers in the UxS industry. We found 49 cases of Chinese UxS manufacturers purchasing from international

---

[45] Nature Index, "2021 Tables: Institutions," webpage, undated. The tables are based on Nature Index data from January 1 to December 31, 2020.

[46] Nurkin, Bedard, Clad, Scott, and Grevatt, 2018.

[47] This has been especially true since Xi Jinping rose to power and his directives have boosted the Chinese Communist Party's influence over private industry; see Lindsay Maizland and Andrew Chatzky, *Huawei: China's Controversial Tech Giant*, New York: Council on Foreign Relations, August 6, 2020.

[48] Kimberly Hsu, *China's Military Unmanned Aerial Vehicle Industry*, Washington, D.C.: U.S.-China Economic and Security Review Commission, 2013.

[49] The full translated quote reads,

> At present, breakthroughs have been made in multisensor environment perception fusion technology, intelligent control technology, and overall UAV technology. However, due to the influence of the national defense technology and industrial base, some key components still rely on imports, and the overall situation has not been free from the passive situation of being constrained by others. (Zhang, Lian, Li, and Yuan, 2020, p. 11)

[50] Charles Clover, "China Seeks Private Sector Help to Streamline Bloated Army," *Financial Times*, February 8, 2015.

[51] Andrew W. Hull, David R. Markov, and Eric Griffin, *"Private" Chinese Aerospace Defense Companies*, Montgomery, Ala.: China Aerospace Studies Institute, Air University, 2021.

suppliers. We researched these suppliers to determine what they produced and where they were headquartered to create a snapshot of some of China's international UxS suppliers. FactSet data only included suppliers for ten of the identified UxS organizations and therefore did not provide an exhaustive list. However, with these caveats, we found that AVIC had 22 foreign suppliers, the State Grid Corporation of China had nine; Ehang Holdings had six; Huawei Technologies had three; and the CAS, China Aerospace Science and Technology Corporation, and DJI had two each. China Electric, Tsinghua Tongfang Company, and Xi'an Aerospace Technology each had one foreign supplier. FactSet also indicated 23 suppliers from Hong Kong, but due to the increasing Chinese influence over Hong Kong, suppliers from Hong Kong have been excluded from the analysis because it is unlikely China would lose access to those supply lines.

Figure 4.6 shows the industrial categories and locations of these foreign suppliers to the Chinese UxS industry, highlighting the dominance of aviation.

Figure 4.6. Industrial Categories of Foreign Suppliers to the Chinese UxS Industry

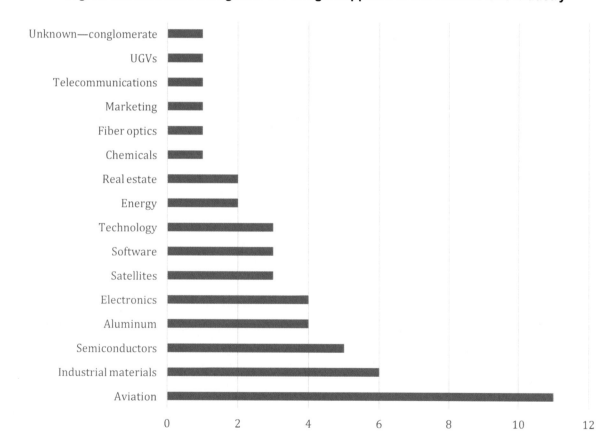

SOURCE: Authors' analysis of FactSet, undated.

The figure indicates that China's largest number of international suppliers associated with the sample set are in the areas of aviation, industrial materials, and semiconductors. A full list of the countries and the supplier's main business is available in Appendix B.

65

Similarly, Figure 4.7 shows the countries where the headquarters of each international supplier is based. The United States has the largest share, with 14 suppliers of Chinese UxS companies, followed by France and Italy, with five suppliers each. These graphs can provide some insight into the foreign dependencies of the Chinese UxS industry but are by no means exhaustive. Additionally, FactSet data do not include the specific items purchased nor the specific UxS platforms on which they are used.

Figure 4.7. Headquarters Locations of Foreign Suppliers to the Chinese UxS Industry

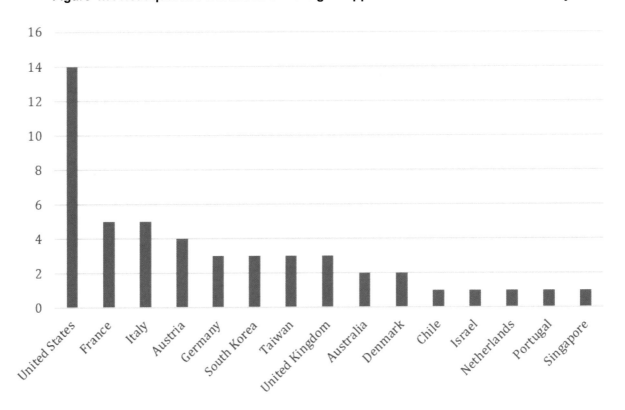

SOURCE: Authors' analysis of FactSet, undated.

Despite sanctions imposed on China in recent years, the nation has become adept at avoiding supply chain disruptions as a result of U.S. sanctions. In 2018, a Chinese state-controlled company bought an Italian manufacturer of military drones. Soon after, it began transferring the company's know-how and technology—which had been used by the Italian military in Afghanistan—to China. The Italian and European authorities had no knowledge of the move.[52]

We also examined data on the number of different Chinese UxS platforms by size, use, and type from the AUVSI database, as shown in Table 4.1. The large number of nondefense UAS

---

[52] Marson and Legorano, 2021.

Table 4.1. Chinese UxS Platform Market Use, by Platform Type

Market Use	UAS	UGS	USV	UUV	Total
Defense only	29	2	3	0	34
Mixed use (including defense)	71	3	1	2	77
Mixed use (excluding defense)	271	96	10	24	401
Total	371	101	14	26	512

SOURCE: Authors' analysis of data from AUVSI, undated a.
NOTES: We excluded inactive platforms and any platforms with insufficient data. We included maritime semisubmersibles with UUVs.

platforms is consistent with China's leadership in the worldwide commercial and recreational UAS market.

Exploring the breakout of the platform market use by size is also interesting, as is shown in Figure 4.8.

Figure 4.8. Composition of Chinese Platforms, by Type and Size

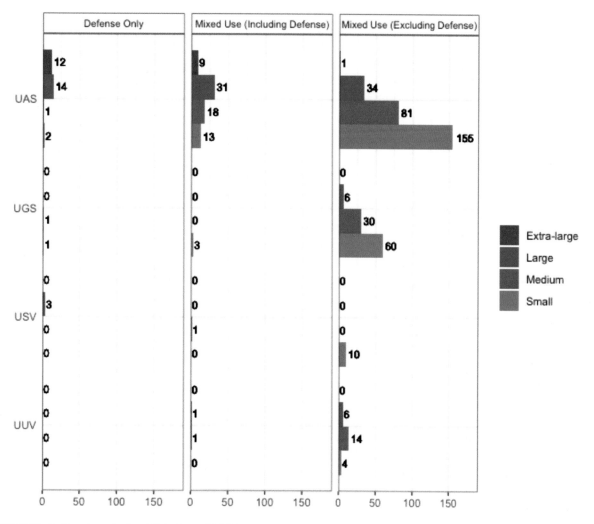

This figure shows that larger platforms tend to be more defense oriented, while smaller platforms tend to have more diversity and are more often used in nondefense applications.

Chapter Summary

From the FaC analysis in this chapter, we conclude that China is concerned about shortcomings in its supply chains and uses government involvement and oversight to direct research and production into areas that interest the PLA.

Second, as noted in the "Interest in International Uncrewed Systems" section, China expects half of U.S. military equipment on the battlefield to be uncrewed by 2040. This shows that China considers the United States to be the primary leader in military UxS development and production, with other competitors being Israel, Russia, and the United Kingdom. China pays very close attention to U.S. UxS and "battle-tested" capabilities across all domains.

In the discussion on platforms, we noted that UAS platforms are an area of Chinese strength, with the company DJI holding approximately 70 percent of the global market for drones. We conclude that Chinese UAS are considered the only domain for short-term practical uses, whereas UGVs, USVs, and UUVs require further development (with Chinese sources citing issues including safety, accuracy, and interoperability issues). Furthermore, China is closely involved in UxS development, due primarily to the nation's military-civil fusion initiative, often funneled through state-run companies and research grants to state universities. While scholars note there are issues in Chinese supply chains, the specifics of these issues are unclear.

Finally, it is worth noting that sanctions will not alone be sufficient to ensure that China does not obtain access to technology and supply lines that it currently lacks, as it has a history of maneuvering around sanctions.

China is ramping up its application of UxS, as noted in many of the Chinese sources, who cite areas where UxS technology lags behind other nations; a desire to reduce risk to soldiers; and domestic applications, such as agriculture and incentives offered to compensate for the impact of the COVID-19 pandemic.[53] Scholars note that key technologies in control, coordination, and pathing will greatly accelerate UxS development in the near term.[54] China is already a leader in UAS development, and it seeks to bring its sea and ground domain vehicles up to U.S. levels through significant military, private company, and research institute coordination.

---

[53] As Zhang, Lian, Li, and Yuan, 2020, p. 14, note,

> Judging from the U.S. military's experience in the development of intelligent unmanned support equipment and [China's] actual situation, we must accelerate the development of intelligent unmanned support equipment to adapt to the future war model and guarantee the goal of winning.

See also Lin and Zhang, 2018.

[54] Liu, Li, and Liang, 2020.

Chapter 5. Assessment of Russian Uncrewed Defense Industrial Base Systems

Russia has both a burgeoning commercial UxS sector and a larger military UxS sector. We reviewed the literature on more than 145 documented Russian platforms in development or production and selected 18 for further analysis.[1] (See Appendix B for the list of sample platforms.)

Demand Signal: Expected Units and Production Cost

We used the Janes Markets Forecast data to examine the expected demand signal for Russian UxS. These data should be taken with a grain of salt due to the lack of transparency associated with data on expected Russian military procurements. Figure 5.1 shows historic and expected Russian UxS units per year.

Like China, Russia appears to have limited numbers of platforms in the pipeline, but expected production of UAS was projected to increase starting in 2024; and there was a large, expected increase in production of UGSs anticipated in 2026. Figure 5.2 shows the USV and UUV platforms at a smaller scale.

We see an expected increase in production of UUVs but limited expected production of USVs. The recent and expected production costs for Russian UxS are shown in Figure 5.3.

Total production costs of Russian UAS were expected to increase sizably, from $85 million in 2021 to $1.1 billion in 2026. However, we did not see corresponding expected increases in production costs for other types of UxS.

---

[1] Systems were compiled from two primary sources: (1) Jeffrey Edmonds, Samuel Bendett, Anya Fink, Mary Chesnut, Dmitry Gorenburg, Michael Kofman, Kasey Stricklin, and Julian Waller, *Artificial Intelligence and Autonomy in Russia*, Arlington, Va.: CNA, May 2021; and (2) Sten Allik, Sean Fahey, Tomas Jermalavičius, Roger McDermott, and Konrad Muzyka, *The Rise of Russia's Military Robots: Theory, Practice and Implications*, Tallinn, Estonia: International Centre for Defence and Security, February 2021. Additionally, system characteristics were corroborated via searching the following sources: *Армейский Стандарт* [*Army Standard*], CNA, *The Drive*, the International Centre for Defense and Security, *Izvestia*, the *Lenta* online newspaper, the *Военное Обозрение* [*Military Review*], the Mil.Press FlotProm website, *Military Today*, RIA Novosti, TASS, the RoboTrends website, and *RT*.

Figure 5.1. Russian Historic and Expected Military UxS Units Delivered per Year, 2016–2026

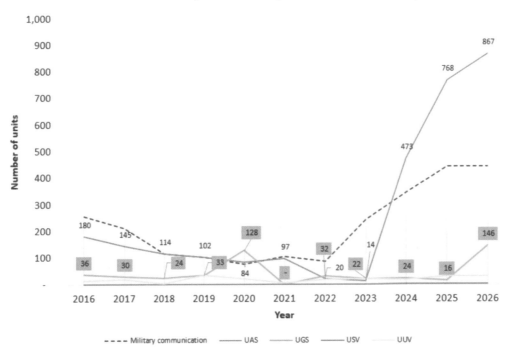

SOURCE: Authors' analysis of data from Janes Markets Forecast.
NOTES: The Military Communication category consists primarily of upgrades to legacy systems. UGS platform counts have an orange background.

Figure 5.2. Russian Historic and Expected Military UxS Units Delivered per Year, 2016–2026: UUVs, USVs, and Military Communication

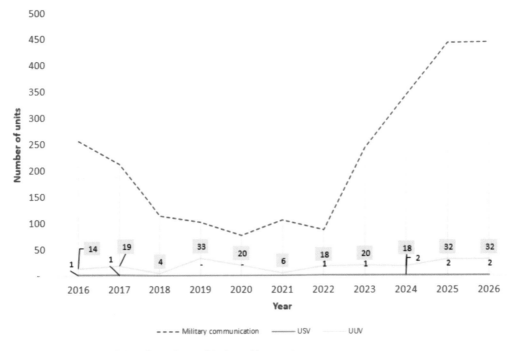

SOURCE: Authors' analysis of data from Janes Markets Forecast.
NOTES: The scale has changed from the previous figure. The Military Communication category consists primarily of upgrades to legacy systems.

Figure 5.3. Russian Expected UxS Production Costs, by Janes Type and Use (in millions of dollars)

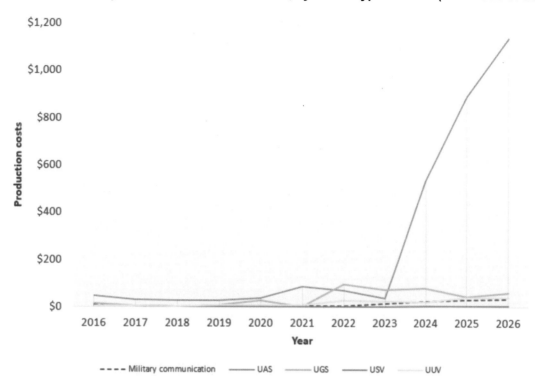

SOURCE: Authors' analysis of data from Janes Markets Forecast.
NOTE: The Military Communication category consists primarily of upgrades to legacy systems.

Exports

While Russia ranks noticeably lower than the United States in terms of uncrewed exports,[2] it nonetheless produces systems that are prominent on the domestic market, and it aspires to grow its export market.

Fragility and Criticality

The sample of 18 Russian platforms chosen for further analysis in this report are analogous to those in the U.S. sample based on size and function. The 12 prime manufacturers in the set produce multiple uncrewed platforms and are active in the Russian uncrewed industrial base. Figure 5.4 demonstrates the extent to which these prime manufacturers (in blue) are connected to one another in the industrial base via associated suppliers, developers, funders, and/or customers (in orange). The figure also reveals the Russian government's relationship with the prime manufacturers: the Russian Ministry of Defense, state-owned Rostec, and funding via the Advanced Research Fund, the Innovation Promotion Fund, and the National Technology Initiative's Russian Venture Capital Fund connect the manufacturers with partner companies and Russian universities. In the figure, entities with more connections are proportionally larger.

---

[2] Janes, "Janes Markets Forecast – UAV, UGV, USV Program Forecast," data as of data as of October 27, 2021.

Figure 5.4. Network of Russian Sample Prime Manufacturers and Associated Entities

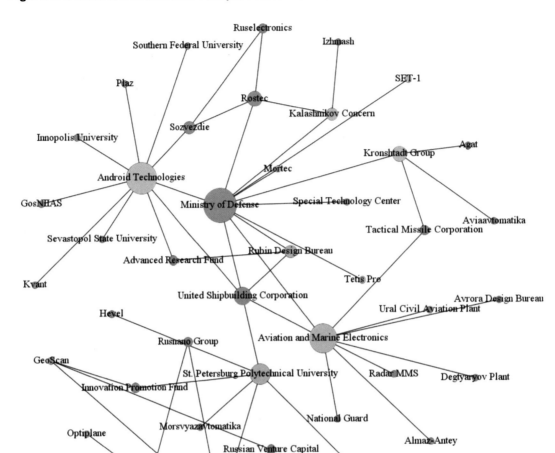

SOURCES: Data from Jeffrey Edmonds, Samuel Bendett, Anya Fink, Mary Chesnut, Dmitry Gorenburg, Michael Kofman, Kasey Stricklin, and Julian Waller, *Artificial Intelligence and Autonomy in Russia*, Arlington, Va.: CNA, May 2021; Sten Allik, Sean Fahey, Tomas Jermalavičius, Roger McDermott, and Konrad Muzyka, *The Rise of Russia's Military Robots: Theory, Practice and Implications*, Tallinn, Estonia: International Centre for Defence and Security, February 2021; Екатерина Гумарова [Ekaterina Gumarova], "Если не станут мешать, через 10 лет российский рынок беспилотников будет не узнать" [If They Don't Interfere, in 10 Years the Russian Drone Market Will Be Unrecognizable], *Реальное Время* [*Real Time*], March 11, 2019; Thomas Newdick and Tyler Rogoway, "Russia's Predator-Style Drone with Big Export Potential Has Launched Its First Missiles," *The Drive*, December 28, 2020; Joseph Trevithick, "This U.S. Army Manual Has New Official Details About the RQ-170 Sentinel Drone," *The Drive*, May 12, 2017; "Russia's State Arms Exporter to Offer Kamikaze Drones, Heavy UAVs to Foreign Customers," TASS, November 25, 2021; Валерий Бутымов [Valery Butymov], "новую морскую модульную платформу FALCO представили на форуме 'Армия-2020'" [The New Marine Modular Platform FALCO was Presented at the Forum "Army-2020"]," *Mil. Press Flot Prom*, August 29, 2020; Вячеслав Прокофьев [Vyacheslav Prokofyev], "РБК: Сколково" и "Роснано" перейдут под управление ВЭБ.РФ [RBC: Skolkovo and Rosnano will come under the control of VEB.RF], *Tass*, November 22, 2020. Rostec, Annual Report of the Rostec State Corporation 2019, Science, Overcoming Technologic Barriers, approved by the Supervisory Board of the Rostec State Corporation April 30, 2020; GeoScan Group, "About Company," undated. Additional data confirmed via press releases of the Russian Advanced Research Fund; however, Russia has since blocked the website to foreign IP addresses due to the War in Ukraine.

The analysis of the sample set of Russian platforms suggests that the majority are primarily composed of Russian components and assemblies, but many still rely on foreign components to some extent. The foreign components most cited in the case studies are cameras and sensors from China and Japan; satellite receivers from China and the United States; batteries and chargers from China; and computer parts, microchips, and engine parts (e.g., starters) from various countries. It should be noted that information on components was not available for all 18 platforms. (For example, little information about components used on the Iskatel USV, produced by Aviation and Marine Electronics, was available outside of conjecture on online forums.) In other cases, we relied on public statements by Russian defense contractors.

Important Russian Firms

In this section we provide a brief description of the main Russian companies involved in the UxS industry and the foreign components used in their products.

GeoScan Group produces the 201 Agrogeo, Lite, Gemini, and the 401 UAS. The company is a commercial manufacturer of fixed- and rotary-wing UAS for mapping and aerial photography to support agriculture, oil and gas development, urban planning, and other industries. The company's production line is based in Russia, and it supplies commercial UAS to 12 countries. The company reports using the following foreign components in its systems:

- Topcon GNSS receivers (United States)
- Sony cameras (China or Japan)
- MicaSense cameras (United States)
- U-blox GNSS receivers (Swiss, production likely in China)
- SkyRC battery chargers (China)
- ImaxRC battery chargers (China)
- iCharger battery chargers (China).[3]

Android Technologies developed the Russian Fedor UGS for operating in dangerous environments with funding from the Advanced Research Fund and a consortium of Russian companies and universities, including Southern Federal University, Innopolis University, and Kvant (a private company). It is working on its next-generation Teledroid.[4] According to press reporting, the executive director of Android Technologies has said that, until 2018, three-fourths of the robot was produced in Russia, while key sensors, motors, cameras, and computer components were imported from abroad (likely from China and Japan).[5]

---

[3] For a detailed list of components, see GeoScan Group, homepage, undated.

[4] Операция «Преемник» идет полным ходом—робота Федора на МКС заменит «Теледроид» [Operation "Successor" Is in Full Swing—Fedor's Robot on the ISS Will Be Replaced by "Teledroid"], *Meduza*, February 8, 2022.

[5] Paulina Glass, "Russia's Pistol-Packing Robot Is Scrambling for Parts," *Defense One*, February 28, 2019.

Tetis Pro produces the Marlin 350 UUV, which has been procured by the Russian Navy since 2018.[6] It could also serve as a dual-purpose UUV for arctic and deep-sea exploration in addition to ISR functions. Janes analysts estimate that 60 percent of the system is produced by Tetis Pro, while about 30 percent of its components come from external, though not necessarily foreign, suppliers. (Ten percent of engine components are of undetermined origin.)[7]

Rubin Design Bureau is a subsidiary of United Shipbuilding Corporation, a large Russian state-owned enterprise. It is responsible for the design and production of the Klavesin-2R-PM, Poseidon, Vityaz-D, and Cephalopod UUVs. These vehicles are produced primarily for the military, and some are still in development; thus, information about their production is limited. (They could also potentially serve as dual-purpose UUVs for deep-water exploration.) Janes analysts estimate that approximately 60 percent of the systems are produced by United Shipbuilding Corporation, while 40 percent of system components are from external, though not necessarily foreign, suppliers.[8]

The Special Technology Center is a Russian defense contractor that produces the Orlan-10 UAS for ISR. At the time of this research, the Special Technology Center also had contracts in place for an Orlan-E export version.[9] Since the Orlan-10 was shot down in combat over Ukraine, information about its foreign components is widely available on the internet. Approximately 60 percent of the platform (its airframe, most of its engine, and many system components) is produced in Russia,[10] and 40 percent of its components have been found to be foreign:

- U-blox GNSS receivers (Swiss, production likely in China)
- STMielectronics flight controller chips (France, Italy)
- Freescale Semiconductor microcircuits/pressure sensors (United States)
- Honeywell compass sensors (United States)
- Texas Instruments engine starters (United States)
- Saito engine ignition modules (Japan)
- Microchip microcontrollers for telemetry transmission modules (United States)
- AnyLink transmitters for telemetry transmission modules (Germany)
- Other associated microchips (China).[11]

---

[6] Janes, "Janes Markets Forecast – UAV, UGV, USV Program Forecast," data as of October 27, 2021.

[7] Janes, "Janes Markets Forecast – UAV, UGV, USV Program Forecast," data as of October 27, 2021.

[8] Janes, "Janes Markets Forecast – UAV, UGV, USV Program Forecast," data as of October 27, 2021.

[9] "First Export of Russian Orlan-E Drones Goes to Myanmar," Defense World.net, January 22, 2021.

[10] Janes, "Janes Markets Forecast – UAV, UGV, USV Program Forecast," data as of October 27, 2021.

[11] "Russian Drone Orlan-10 Consists of Parts Produced in the USA and Other Countries—Photo Evidence," Inform Napalm, February 6, 2018.

Kronshtadt Group is a defense contractor owned by the large Russian conglomerate Sistema; it produces the Orion UAS. At the time of this research, the company was working with Rosoboronexport (the state-owned arms exporter) to export combat and ISR versions of the platform.[12] According to media statements from Kronshtadt, the Orion system relies solely on Russian components, materials, and assemblies.[13] Janes analysts estimate that 70 percent of the system—including the airframe and major system components—is produced by Kronshtadt, while the remaining 30 percent is likely sourced from external, though not necessarily foreign, suppliers.[14] This likely includes an engine from the Russian company Agat, which replaced a failed engine from another Russian company, Itlan.[15]

SET-1 is a Russian defense contractor that produces the Scarab and Scorpion (a next-generation prototype) UGVs for bomb detection and ISR. Both are smaller, four-wheel-drive, track-optional models weighing under 17 kg that can be easily transported to combat and can operate in extreme climates. SET-1 states that it has in-house R&D, engineering, and manufacturing departments that employ more than 70 engineers.[16] Janes analysts estimate that 60 percent of the system's components are produced by SET-1 and an additional 40 percent may be produced by SET-1 or external, though not necessarily foreign, suppliers.[17]

Commercial and Defense Sales

Many large Russian state-owned companies do not publish data on revenues or sales of UxS separately in annual reports (e.g., Rostec's Kalashnikov Concern, Sistema's Kronshtadt Group). Furthermore, data from FactSet are limited to publicly traded companies, few of which are heavily involved in the UxS industrial base. While publicly traded companies, like KAMAZ, are connected to universities that develop uncrewed platforms, the FactSet data on their foreign suppliers were not specific enough to identify trends for the industry. Thus, we ascertained aspects of FaC from a review of the literature on Russian UxS and the DIB. Trends toward more Russian self-sufficiency in critical sectors, such as defense and shipbuilding, have evolved since 2016, when Prime Minister Dmitry Rogozin lamented the embezzlement of state funds and

---

[12] Inder Singh Bisht, "Russia Offers Orion-E Combat Drone for Export," *Defense Post*, June 29, 2021.

[13] Алексей Заквасин [Zakvasin, Aleksei] and Елизавета Комарова [Elizaveta Komarova], "«Технологическая состоятельность»: руководители «Кронштадта» — о возможностях БПЛА «Орион» и планах по развитию компании" ["Technological Viability": The Leaders of "Kronstadt"—About the Capabilities of the UAV "Orion" and Plans for the Development of the Company], *RT*, October 10, 2021.

[14] Janes, "Janes Markets Forecast – UAV, UGV, USV Program Forecast," data as of October 27, 2021.

[15] Newdick and Rogoway, 2020.

[16] Information about SET-1 was taken from SET-1, "About Us," webpage, undated.

[17] Janes, "Janes Markets Forecast – UAV, UGV, USV Program Forecast," data as of October 27, 2021.

Admiral Viktor Chirkov of the Russian Navy criticized an overreliance on foreign-made parts.[18] However, concerns still exist, and while Russia does have an import substitution program mandating the use of domestically made components (the Forpost-R and Altius-U UAS are manufactured by the Ural Civil Aviation Factory and have been affected by this program, for example),[19] even in the Soviet Union, not all military equipment was domestically made. A major obstacle is that a significant portion of Russia's machine tools used in plants and factories are, themselves, imported.[20] This trend is similar across the chemical, printing, and textile sectors, which import 60–80 percent of equipment needed for production.[21] Sanctions may have further exacerbated this problem.

Despite these issues, Russia still has both a burgeoning commercial UxS sector and a larger military UxS sector. In 2018 and 2019, Russian president Vladimir Putin launched national programs to invest in Russia's digital economy, including UxS and AI, through 2030.[22] As a result, a significant portion of Russia's commercial UxS sector benefits from some state funding or management via a state-owned enterprise. The following are examples of how state support has reached the commercial sector:

- SKYF Agro, a large commercial UAS in development from ARDN Technology, a startup, has been supported by the National Technology Initiative's Russian Venture Capital Fund.[23]
- Cognitive Pilot, which is developing AI for autonomous vehicles, is a venture of SberBank, one of Russia's largest state-owned enterprises.[24]
- KAMAZ, a publicly traded company with some state ownership via Rostec, is working with Innopolis University on uncrewed trucks.[25]

Thus, Russia has a strong uncrewed industrial base that includes manufacturing resources (e.g., plants, factories, shipyards, and machine shops), with the acknowledgment that the plants likely rely on foreign inputs. Additionally, the country has a well-educated populace and strong technical universities (e.g., the Skolkovo Institute, Innopolis University). Higher education is

---

[18] Vladimir Voronov, *Import Substitution for Rogozin*, trans. Arch Tait, London: Henry Jackson Society, January 2016.

[19] Mönch Publishing Group, "Russia to Receive New-Generation UAS by 2030," February 25, 2020.

[20] Voronov, 2016.

[21] Paul Goble, "Import Substitution in Russia Failing as Moscow Buys Products Not Technologies," *Eurasia Daily Monitor*, Vol. 16, No. 44, March 28, 2019.

[22] Edmonds, Bendett, Fink, Chesnut, Gorenburg, Kofman, Stricklin, and Waller, 2021.

[23] Gumarova, 2019.

[24] Information about Cognitive Pilot was taken from SberBank, "Cognitive Pilot," webpage, undated.

[25] Information about the project was taken from the university's website; see Innopolis University, "Innopolis University Develops and Autonomous Driving System for KAMAZ," press release, Innopolis, Russia, February 8, 2018.

both a legacy of the Soviet Union and the result of more recent investments under Putin. However, regarding human capital, there remains a risk of a "brain drain" of young, educated talent, as observed in recent polls, and following the beginning of the war with Ukraine in 2022.[26] Furthermore, Russia has abundant natural resources, with iron and steel being among its top exports. The country is also a prominent global supplier of aluminum, nickel, copper, titanium, and other nonferrous metals that are important to the UxS industrial base.[27] It should also be noted that the Russian federal budget is heavily reliant on hydrocarbon exports for revenues, and, despite a decade of high oil and gas prices, Russian spending on R&D (acknowledging limitations for unavailable and/or classified data) has remained stubbornly close to only 1 percent of its gross domestic product, while China's share grew from 1.5 to 2.2 percent and the U.S. share averaged 2.7 percent over the same period (2008–2018).[28]

Defense Uniqueness

We reviewed the AUVSI data on the Russian platform market mix, as shown in Table 5.1. Russia's UxS platforms tend to be more defense oriented; there are relatively few nondefense platforms in comparison with the United States and China.

Table 5.1. Russian UxS Platform Market Use, by Platform Type

Market Use	UAS	UGS	USV	UUV	Total
Defense only	18	3	0	0	21
Mixed use (including defense)	61	5	0	4	70
Mixed use (excluding defense)	13	5	0	6	24
Total	92	13	0	10	115

SOURCE: Authors' analysis of data from AUVSI, undated a.
NOTES: We excluded inactive platforms and any platforms with insufficient data. We included maritime semisubmersibles with UUVs.

We also looked at the AUVSI data by size, as shown in Figure 5.5.

The figure shows that larger platforms tend to be more defense oriented, whereas smaller platforms tend to have more diversity and use in nondefense applications. However, even small and medium UAS platforms are more likely to include some defense uses rather than being purely nondefense in nature.

---

[26] Levada-Center, "Emigration," July 6, 2021.

[27] V. K. Fal'tsman, "Import Substitution in the Economic Sectors of Russia," *Studies on Russian Economic Development*, Vol. 26, No. 5, 2015.

[28] World Bank, "Research and Development Expenditure (% of GDP)," graph, data as of June 2022.

Figure 5.5. The Composition of Russian Platforms, by Type and Size

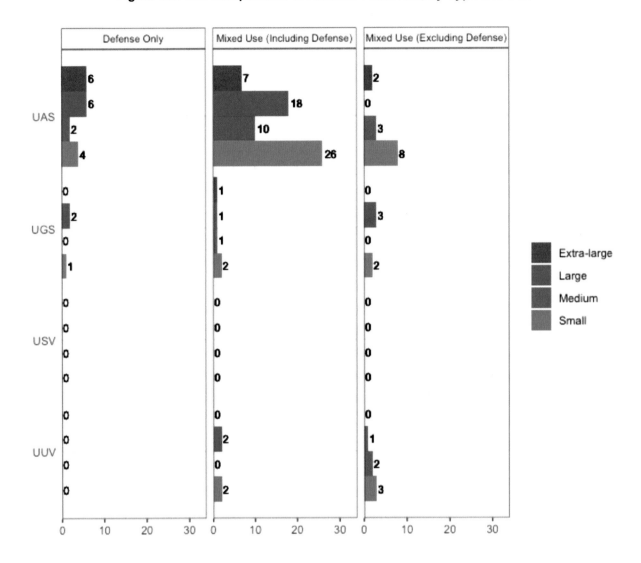

SOURCE: Authors' analysis of data from AUVSI.

Conclusions

For its part, the Russian Ministry of Defense has pursued the development of uncrewed platforms as a logical step since its modernization began over a decade ago with the support of a much stronger Russian federal budget and more ambitious state procurement plans. The Russian Ground Forces (the nation's army) have successfully integrated UAS into combat in Ukraine, including the Granat (produced by the Kalashnikov Concern, a subsidiary of the state-owned enterprise Rostec) and the Orlan-10 (produced by the Special Technology Center).[29] A drone attack on Russian forces by independent actors in Syria also contributed to plans for more

---

[29] Allik, Fahey, Jermalavičius, McDermott, and Muzyka, 2021.

78

anti-UAS capabilities.[30] In terms of UGSs, in addition to the combat-tested Uran-6 in Syria (Rostec) and the Uran-9 under development, a consortium of the Advanced Research Fund, Android Technologies, Southern Federal University, and private companies is developing a potentially AI-enabled combat UGV called Marker. One goal of future applications like Marker is the integration of UAS and UGV platforms. Finally, the state-owned United Shipbuilding Corporation is developing and producing UUVs for the Russian Navy that can be constantly submerged for both defensive and offensive capabilities.[31] One example is the Poseidon, an offensive UUV whose nuclear torpedoes have been touted as having the ability to generate a radioactive tsunami-style attack on the U.S. coast.[32]

A primary risk for U.S. and Western forces is likely Russia's intent to incorporate AI into its uncrewed platforms for use across combat domains. The Russian Ministry of Defense has already experimented with UAS platforms for electronic warfare and C2. It is believed that Russian military planners see the next evolution of these steps being combined "swarms" of UAS supporting crewed aerial vehicles in combat situations to improve situational awareness and enhance electronic warfare capabilities to disrupt communications.[33] While UGS use on the battlefield may still be many years away, the Russian military is also experimenting—specifically, with the combined use of UAS and UGSs, as observed in the military exercise Zapad-21. The synthesis of these platforms could improve situational awareness and C2 in combat. Finally, the Russian Navy is experimenting with UUVs that could challenge U.S. superiority at sea by forcing the United States to refocus on less costly robotic adversaries with strong situational awareness.[34] Russia's Poseidon has caused particular concern that the West does not have anything yet that can travel deep and fast enough to successfully counter such a UUV.[35]

Taken together, the main issue seems to be Russia's intention to counter U.S. and Western superiority asymmetrically across multiple domains with less costly, AI-enabled uncrewed platforms that do not put Russian personnel in harm's way. Furthermore, as some analysts have noted, the Russian military may have fewer reservations about the dangers of fully AI-enabled weapons than North Atlantic Treaty Organization (NATO) members.[36]

---

[30] David Reid, "A Swarm of Armed Drones Attacked a Russian Military Base in Syria," CNBC, January 11, 2018.

[31] Allik, Fahey, Jermalavičius, McDermott, and Muzyka, 2021.

[32] Michael Peck, "This Massive Russian Sub Is Preparing to Launch Nuclear Torpedoes," *National Interest*, September 17, 2021b.

[33] Allik, Fahey, Jermalavičius, McDermott, and Muzyka, 2021.

[34] Edmonds, Bendett, Fink, Chesnut, Gorenburg, Kofman, Stricklin, and Waller, 2021.

[35] Mark Episkopos, "The Russian Navy Loves Drones (And for Good Reason)," *National Interest*, June 22, 2021.

[36] Allik, Fahey, Jermalavičius, McDermott, and Muzyka, 2021.

There are opportunities to counter Russian uncrewed platforms. First, as we have noted, the Russian UxS industrial base seems to lack key assemblies such as sophisticated electronics and engine components, and it has a heavy reliance on China. Furthermore, machine shops and plants may still rely heavily on foreign-built machine tools. As a result of Russia's invasion of Ukraine in 2022, the United States will restrict technological exports to Russia, such as semiconductors, aviation parts, lasers, sensors, and technologies produced in foreign countries (e.g., Taiwan) that rely on U.S.-developed software and equipment.[37] Although the United States exported just $114 million in microchips to Russia in 2021, and Russia will likely rely on China for substitution, the lack of access to these products will negatively affect the Russian UxS industrial base over time.[38] Semiconductor (and subsequently microchip) supply chains will likely become further disrupted due to Ukraine being a key player in the production of neon.[39] The United States should therefore monitor Russia's import substitution from China while acknowledging that Russia's more restricted ability to work with outsiders (as intended by sanctions) could make tracing and monitoring developments in the industry all the more difficult.

Second, many of the military platforms discussed in this chapter rely heavily on Russian military infrastructure (i.e., command, control, communications, computers, intelligence, surveillance, and reconnaissance), and analysts suggest that countermeasures to disrupt this infrastructure are better investments than pursuing one-to-one UxS parity with Russia.[40] In general, Russia has demonstrated advancements in its uncrewed technologies. And while Russia may not be a major exporter of uncrewed platforms at the moment, it remains the second-largest exporter of arms after the United States.[41] At the time of this research, Russia had a few contracts with Myanmar and unspecified countries for UAS exports, and it was not clear how Russia's invasion of Ukraine would affect future contracts.[42] Notably, the deal with Myanmar was for the Orlan-10E export version, and, as previously reported, many foreign components were discovered on the Orlan-10.[43] Thus, in the short term, the largest procurer of Russian uncrewed platforms will remain the Russian Ministry of Defense. It is possible that the top-down government approach to funding R&D may mean that defense-related projects are overly dominated by

---

[37] White House, "Fact Sheet: Joined by Allies and Partners, the United States Imposes Devastating Costs on Russia," February 24, 2022.

[38] Tim Culpan and Tae Kim, "Tech Sanctions Won't Sting Russia for a While," Bloomberg, February 25, 2022.

[39] Meaker, Morgan. "Russia's War in Ukraine Could Spur Another Global Chip Shortage," *Wired*, February 28, 2022.

[40] Edmonds, Bendett, Fink, Chesnut, Gorenburg, Kofman, Stricklin, and Waller, 2021.

[41] SIPRI, "Top List TIV Tables," web database, undated, data as of February 10, 2022.

[42] "Russia's State Arms Exporter," 2021.

[43] David Hambling, "Russia Enters Military Drone Export Market with Sale to Myanmar," *Forbes*, January 25, 2021a.

Russian state-owned enterprises, which could limit the resources available to smaller and medium-size enterprises, stunt innovation, and create inefficiencies in the commercial sector.[44] Still, given Russia's efforts to modernize its military technology and experiment with more autonomy in UxS to create asymmetric challenges to the United States and the West, Russia's status as a competitor should not be underestimated. Further monitoring of its uncrewed industrial base, and how the industry will navigate the unprecedented sanctions from the war in Ukraine, is warranted.

---

[44] Edmonds, Bendett, Fink, Chesnut, Gorenburg, Kofman, Stricklin, and Waller, 2021.

Chapter 6. The Competitiveness of U.S. Exports of Uncrewed Systems

In this chapter we examine the competitiveness of U.S. exports of UxS against those produced by other countries. First, we review U.S. export regulations affecting UAS, UGSs, and uncrewed maritime systems, including both USVs and UUVs. Next, we examine patterns in global defense exports of UxS using historical Janes data for the period 2009–2020. Finally, we identify risks, considerations, and opportunities for DoD.

U.S. Export Controls on Uncrewed Systems

UAS face the tightest restrictions on foreign sales. UAS with a range greater than 300 km and a payload capacity greater than 500 kg are classified as Category 1 in the Missile Technology Control Regime, a multilateral treaty intended to limit the proliferation of missile technology. Category 1 items are subject to a strong presumption of denial regardless of the purpose of the export and are licensed for export only on rare occasions.[1]

International Traffic in Arms Regulations applies to UAS, UGSs, USVs, and UUVs on the United States Munitions List (USML).[2] These systems can only be sold through Foreign Military Sales (from the U.S. government to foreign governments) or with an export license issued by the U.S. Department of State.

UAS *not* covered by the Missile Technology Control Regime *are* covered by the USML if they function as bombers, fighters, fixed-wing attack aircraft, or attack helicopters; incorporate defense articles that perform ISR, electronic warfare, airborne warning and control, or command, control, and communications (C3) functions; or are armed or specially designed to deliver munitions or otherwise destroy targets. The USML also includes target drones, airborne launching systems for UAS, launching and recovery equipment designed to allow a UAS to take off or land on a vessel, developmental aircraft funded by DoD, threat-adaptive autonomous flight control systems, and UAS flight control and vehicle management systems with swarming capability.

USVs and UUVs are covered by the USML if they meet one of the following criteria:

- they are considered combatant vessels (including mine sweepers, mine hunters, and other mine countermeasures), submarines designed for military use, or antisubmarine warfare vehicles

---

[1] International Trade Administration, "Unmanned Aircraft Systems," webpage, undated.

[2] Code of Federal Regulations, Title 22, Chapter 1, Subchapter M, Part 121, The United States Munitions List, last amended June 30, 2022.

- they are armed or designed to deliver munitions or otherwise destroy or incapacitate targets
- they incorporate mission systems that perform military functions, such as military communication, electronic warfare, target designation, surveillance, target detection, or sensor capabilities
- they are developmental vessels funded by DoD.

In addition, the USML includes control and monitoring systems for autonomous uncrewed vessels capable of onboard, autonomous perception and decisionmaking necessary for the vessel to navigate without human intervention.

Armored ground combat vehicles (including UGSs) are covered by the USML if they meet one of the following criteria:

- they are armed or designed to be used as a firing or launch platforms to deliver munitions or otherwise destroy or incapacitate targets
- they incorporate mission systems that perform military functions
- they are capable of off-road or amphibious use and specially designed to transport or deploy personnel or materiel or to move with other vehicles over land in close support of combat vehicles or troops.

In addition, the USML includes kits that are specially designed to convert these types of vehicles into either uncrewed or driver-optional vehicles and provide remote or autonomous steering, acceleration and braking, and a control system.

The USML also includes any part, component, accessory, attachment, equipment, or system that is classified, contains classified software, or is being developed using classified information.

Other UAS, UGSs, USVs, or UUVs specially designed for military applications or that have the potential for military and civilian applications may require an export license from the U.S. Department of Commerce if they are listed in the Export Administration Regulations, also known as the Commerce Control List (CCL).[3] The regulations provide a reason for control (such as antiterrorism efforts or national security) that determines which end user countries require export licenses. Items not included on the CCL are controlled only for Crimea, Cuba, North Korea, and Syria.[4]

---

[3] U.S. Department of Commerce, Bureau of Industry and Security, Export Administration Regulations, Supplement No. 1 to Part 774, Category 8—Marine, September 11, 2020; Category 9—Aerospace and Propulsion, March 29, 2021; and Category 0—Nuclear Materials, Facilities, and Equipment (and Miscellaneous Items), February 3, 2022, apply to USVs and UUVs, UAS, and UGSs, respectively.

[4] Anthony Rapa, "From Unmanned Systems Magazine: Unmanned Systems and Export Controls: What Your Company Needs to Know," Association for Unmanned Vehicle Systems International, May 21, 2018.

UAS are covered by the CCL if they have the following characteristics:

- they are designed to have controlled flight out of the direct line of sight of the operator and either
  - a maximum endurance greater than or equal to 30 minutes but less than 1 hour and designed to take off and have stable, controlled flight in wind gusts greater than or equal to 25 knots, or a maximum endurance of 1 hour or greater, or
  - incorporate an aerosol dispensing system with a capacity greater than 20 liters
- they have engines specially designed or modified to operate at altitudes greater than 50,000 feet
- they have autonomous flight control or navigation capability
- they are specially designed for any military use not enumerated in the USML.[5]

Uncrewed submersible vehicles are covered by the CCL if they have any of the following characteristics:

- they are designed for deciding a course relative to any geographical reference without a real-time human assistant
- they employ acoustic data or command links
- they employ optical data or command links exceeding 1,000 m
- they are designed to operate with a tether at depths exceeding 1,000 m and are either designed for self-propelled maneuver using certain types of propulsion motors or thrusters or a fiber-optic data link
- they are robots specially designed for underwater use, controlled using a dedicated computer, and having any of the following:
  - control systems using information from sensors that measure force or torque applied to an external object, distance to an external object, or tactile sense between the robot and an external object
  - the ability to exert a force of 250 N or more or a torque of 250 Nm[6] or more and use titanium-based alloys or composite materials in their structural members
  - remotely controlled articulated manipulators with control systems using sensors that measure torque or force applied to an external objects or tactile sense between the manipulator and the object; or proportional master-slave techniques having five or more degrees of freedom of movement.[7]

UGVs are covered by the CCL if they are specially designed for a military use and not enumerated in the USML. These include unarmored military recovery and other support vehicles, and unarmored, unarmed vehicles with mounts or hard points for firearms of .50 caliber or less. These vehicles are considered to be specially designed for military use if they include components

---

[5] U.S. Department of Commerce, Bureau of Industry and Security, 2021.

[6] A newton (N) is the force needed to accelerate one kilogram of mass at the rate of one meter per second squared in the direction of the applied force. A newton meter (Nm) is equal to the torque resulting from a force of 1 N applied perpendicularly to a moment arm that is 1 m long.

[7] U.S. Department of Commerce, Bureau of Industry and Security, 2020.

such as pneumatic tire cases specially designed to be bulletproof, armored protection of vital parts such as fuel tanks or vehicle cabs, special reinforcements or mountings for weapons, or blackout lighting. In addition, unarmed vehicles that are derived from civilian vehicles are covered by the CCL if they are manufactured or fitted with ballistic protection equal to or better than level III, have a transmission that provides drive to both front and rear wheels simultaneously or additional wheels for load-bearing purposes, gross vehicle weight rating greater than 4,500 kg, and designed or modified for off-road use.[8]

In the 2010s, U.S. concerns about proliferation of military drones sold by China and other countries led to the relaxation of some export controls by the administration of President Barack Obama in February 2015 and by the administration of President Donald Trump in April 2018.[9] The 2018 policy change allowed transfers of armed UAS to be made via foreign military sales or direct commercial sales with the restriction that they not be armed with a foreign system or currently unauthorized U.S. system without prior U.S. government authorization. It also allowed transfers of unarmed UAS with the restriction that they not be armed with U.S. or foreign equipment without U.S. government permission.[10]

As noted in our discussion of the Chinese UxS DIB in Chapter 4, in March 2020 SIPRI announced that in 2014–2018, China became the largest exporter of uncrewed combat aerial vehicles. The report also noted that China is not a signatory to the Missile Technology Control Regime, and it imposes relatively fewer restrictions on how buyers use the weapons they import, even if doing so violates international law and human rights.[11] We examined SIPRI's database and found that it indicates that the United States exported 291 ISR drones and 36 combat drones from 2010 to 2020, while China exported 220 combat drones and 18 ISR drones. We conducted a global assessment of UxS exports based on Janes UxS data, which we describe in the next section. This assessment indicates that the United States was the largest exporter of defense UAS from 2009 to 2020 by dollar value, while China ranked fifth.[12]

---

[8] U.S. Department of Commerce, Bureau of Industry and Security, 2022.

[9] Dee Ann Divis, "Enthusiasm Builds for U.S. Military Drone Exports After Rule Change," *Inside Unmanned Systems*, July 11, 2019; U.S. Department of State, "U.S. Policy on the Export of Unmanned Aerial Systems," fact sheet, May 21, 2019.

[10] U.S. Department of State, 2019.

[11] Snehesh Alex Philip, "China Has Become a Major Exporter of Armed Drones, Pakistan Is Among Its 11 Customers," ThePrint, November 23, 2020; Michael Peck, "Coming Soon: World-Class Chinese Military Drones," *National Interest*, July 10, 2021a.

[12] However, a comparison of the SIPRI and Janes data for Chinese UAS exports indicated that Janes did not capture as many sales of CH-3, CH-4, and Wing Loong UAS as SIPRI, although Janes captured military sales of DJI UAS that were not included in the SIPRI data.

Global Patterns in Exports of Defense Uncrewed Systems

We examined Janes data on historical sales of defense UxS over the period 2009–2020 to identify the top exporters of UAS, UGSs, USVs, and UUVs. In this section we describe the results of that analysis. Unfortunately, comparable data are not readily available for commercial and recreational sales of UxS.

Uncrewed Aerial Systems

Figure 6.1 provides an overview of global patterns in exports of defense UAS from 2009 through 2020, focusing on the top ten exporting countries and the end user countries that purchased from them. The size of the country node correlates with the dollar value of exports, with the top two exporters (the United States in red, and Israel in blue) having the largest nodes. Clockwise connecting lines show the flow of exports from the country of manufacture. This market was by far the largest among the three types of UxS, with total defense exports of $9.5 billion over the period.

Figure 6.1. Exports of Defense UAS by Dollar Value, 2009–2020

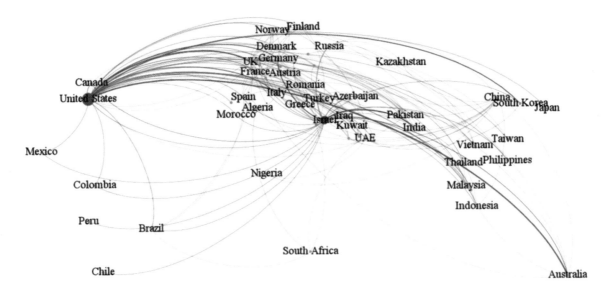

SOURCE: Authors' analysis of data from Janes Markets Forecast.
NOTES: Exports from Israel are in blue, and the United States in red; other countries are shown in gray.
UAE = United Arab Emirates; UK = United Kingdom.

A summary of the dollar value and market share for the top ten exporting countries is provided in Table 6.1. Note that Russia ranked eleventh, with $46 million in exports of defense UAS from 2009 to 2020 and less than 1 percent market share. Additional details on the top exporting companies, system names, and end user countries for the top five exporting countries are listed in Appendix A. We also explored the dollar value of exports by country, as shown in Table 6.1.[13]

Table 6.1. Dollar Value of Sales and Market Share for Top Ten Exporters of UAS

Rank	Country of Manufacture	Dollar Value of Sales, 2009–2020 ($ millions)	Market Share (%)
1	United States	4,579	48
2	Israel	2,558	27
3	Italy	743	8
4	France	504	5
5	China	195	2
6	United Kingdom	182	2
7	Germany	152	2
8	Austria	123	1
9	Turkey	100	1
10	Norway	87	1

SOURCE: Janes Markets Forecast.

Uncrewed Surface Vehicles and Uncrewed Underwater Vehicles

Figure 6.2 provides similar information on global patterns in exports of defense USVs and UUVs for 2009–2020, focusing on the top ten exporting countries and the end user countries that purchased from them. As in the previous figure, the size of the node indicates the dollar value of exports; in this case the top three exporters are Sweden (yellow), France (blue), and the United States (red). Clockwise connecting lines show the flow of exports from the country of manufacture. In comparison with the United States, where exports only accounted for 3 percent of USV and UUV sales, European producers were more dependent on export sales, which accounted for 59 percent of sales in Sweden, 48 percent in France, 47 percent in Norway, and 100 percent in Iceland. (See Appendix A for additional details.)

---

[13] Note that even prior to the 2022 conflict in Ukraine, Turkish UAS exports were growing due to their successful use in Libya, Nagorno-Karabakh, and Syria. Some countries have turned to Turkey because they were unable to get permission to buy U.S. technology, found U.S. systems too expensive, or did not want to import from China or Israel. See "Export Successes Are Vindicating Turkey's UAS Drive," *Aviation Week*, June 8, 2021; and Rick Rozoff, "Turkey's Arms Sales to Ukraine up 30 Times in First Quarter over Last Year," *Daily Sabah*, April 6, 2022.

Figure 6.2. Exports of Defense USVs and UUVs by Dollar Value, 2009–2020

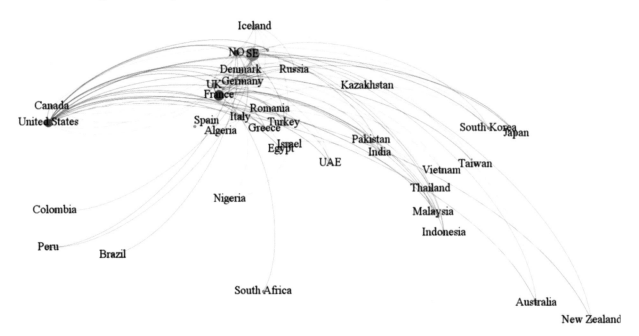

SOURCE: Authors' analysis of data from Janes Markets Forecast.
NOTES: Exports from France are shown in blue, Iceland in green, Norway in magenta, Sweden in yellow, and the United States in red. NO = Norway; SE = Sweden; UAE = United Arab Emirates; UK = United Kingdom.

The defense USV and UUV export market is much smaller than the UAS market, with total export sales of $937 million over the period. A summary of the dollar value and market share for the top ten exporting countries is provided in Table 6.2.

Table 6.2. Dollar Value of Sales and Market Share for Top Ten Exporters of USVs and UUVs

Rank	Country of Manufacture	Dollar Value of Sales, 2009–2020 ($ millions)	Market Share (%)
1	Sweden	258	28
2	France	200	21
3	United States	130	14
4	Norway	84	9
5	Iceland	68	7
6	United Kingdom	66	7
7	Italy	48	5
8	Israel	36	4
9	Germany	10	1
10	Canada	9	1

SOURCE: Janes Markets Forecast.

Note that Janes did not list any exports of defense USVs and UUVs from either China (or to China) or Russia during this period. Additional details on the top exporting companies, system names, and end user countries for the top five exporting countries are provided in Appendix A.

Uncrewed Ground Systems

Figure 6.3 shows similar data on global patterns in exports of defense UGSs from 2009 through 2020, focusing on the top ten exporting countries and the end user countries that purchased from them. The size of the country node indicates the dollar value of exports; in this case, the United States (in red) is by far the largest exporter, accounting for more than half of defense UGS export sales. Clockwise connecting lines show the flow of exports from the country of manufacture.

Figure 6.3. Exports of Defense UGSs by Dollar Value, 2009–2020

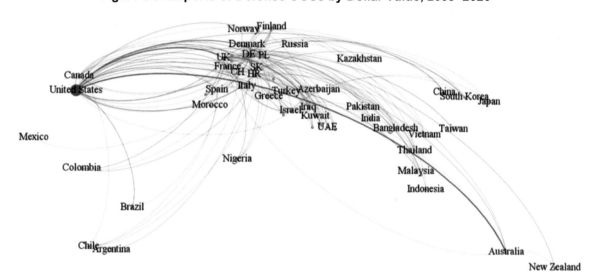

SOURCE: Authors' analysis of data from Janes Markets Forecast.
NOTES: Exports from Croatia are shown in orange, Germany in green, the United Kingdom in blue, and the United States in red. CH = Switzerland; DE = Germany; HR = Croatia; PL = Poland; SK = Slovakia; UAE = United Arab Emirates; UK = United Kingdom.

This export market was a little over twice the size of the USV/UUV market, with total export sales of just under $2 billion during the period. A summary of the dollar value and market share for the top ten exporting countries is provided in Table 6.3.

Note that Russia ranked twenty-first out of 22 exporters, with $508,000 in exports of defense UGSs from 2009 to 2020, and Janes did not record any Chinese exports of UGSs. Exports from Ireland were to an unidentified country and not included in the figure. Additional details on the top exporting companies, system names, and end user countries for the top five exporting countries are provided in Appendix A.

Thus, despite the export restrictions discussed above, the United States is the largest exporter of defense UAS and UGSs, and it ranks third in exports of USVs and UUVs.

Table 6.3. Dollar Value of Sales and Market Share for Top Ten Exporters of UGSs

Rank	Country of Manufacture	Dollar Value of Sales, 2009–2020 ($ millions)	Market Share (%)
1	United States	1,036	56
2	Germany	359	20
3	Croatia	88	5
4	United Kingdom	79	4
5	Israel	60	3
6	Poland	52	3
7	Slovakia	48	3
8	Switzerland	23	1
9	Ireland	22	1
10	Denmark	14	1

SOURCE: Janes Markets Forecast.

Conclusions

Given the large volumes of exports, particularly UAS, proliferation of UAS technologies to countries that are current or potential adversaries is a risk for DoD and could warrant additional investment in counter-UAS technologies. Although additional U.S. exports of UxS could potentially improve the financial health of U.S. manufacturers and help ensure that allies have advanced capabilities, the United States is already a leading exporter of defense UxS and must also consider proliferation risks.

Concerns about the use of Chinese components might have had a greater effect on commercial and recreational sales and exports of U.S.-manufactured UxS than on defense sales and exports, though we do not have access to comparable data on commercial UxS exports to examine this hypothesis. Efforts by the DIU to identify and certify acceptable components would likely help U.S.-based manufacturers, but their products would probably still be more expensive than those produced by China. While the costs of U.S.-made UxS will almost always be higher than those made in China, perceptions of greater quality, capability, and reliability of U.S. systems may be more important in defense than commercial markets.

Chapter 7. Conclusions and Recommendations

This chapter discusses overall conclusions and provides comparative analysis of the U.S., Russian, and Chinese DIBs based where applicable. We also provide a set of recommendations and potential next steps for OUSD(R&E).

Conclusions

U.S. Demand Signal

We found that the overall trend in U.S. demand was increasing for all types of UxS, though the scale of the increase varied significantly by platform type. Only UAS unit demand was expected to increase more than 100 percent from FY 2021 levels, driven by demand for small, low-cost platforms from FY 2024 to FY 2026. UGS unit demand increased substantially from FY 2019 to FY 2021 and was expected to increase 31 percent from FY 2021 to FY 2026, with demand peaking in FY 2024, but the cost growth over the same period was projected to be much larger, at 614 percent. Expected USV and UUV unit demand was essentially flat from FY 2021 to FY 2026; this still represented an increase over historical demand levels.

Overall, estimated UxS procurement in FY 2021 ranged from $1.6 billion (based on Janes) to $3.3 billion (based on AUVSI) in the United States. FY 2026 costs could range from $4.8 billion to $8.4 billion, though there is likely significant uncertainty in these estimates.

Relative to limited commercial data on sales volumes, the DoD demand signal is low. DoD demand in 2022 ranged from 1,200 to approximately 3,000 (based on the trend line) units, suggesting that DoD UAS demand accounts for roughly 1–3 percent of overall yearly U.S. commercial demand. DoD demand accounts for a larger share of the UAS market in terms of dollar value, though the exact percentage is difficult to estimate due to uncertainty about the exact value of both DoD and commercial sales. Estimates of DoD procurement spending range from $1.4 billion to $3.3 billion, while estimates of North American commercial sales range from $4 billion to $11 billion.

U.S. Defense Industrial Base Capacity Criticality

Defense-Unique Capability

A mix of commercial demand and use implies a healthy market for UAS and UGS, with the UUV market smaller relative to those for UAS and UGSs. The USV market is small compared with the others, and while it may not be defense specific, it has few platforms overall.

Defense Design Requirements

This is a challenging area for DoD; a broad array of defense policy requirements hinder the rapid use of commercial platforms for defense use cases, and efforts to source more from U.S. first-, second-, and third-tier suppliers are making it more challenging. However, efforts like DIU's Blue UAS project (focused on small UAS) are intended to combat these challenges by preapproving commercial platforms for use across DoD and the U.S. government.

Skilled Labor Requirements

We view this as a moderate problem area, but we were unable to draw comprehensive quantitative findings. Vendors and government program offices alike identified issues with finding and retaining STEM professionals, such as AI and cybersecurity specialists, as well as skilled labor with UxS experience in such fields as high-end machining and welding. However, none of them indicated that this was a particularly acute problem area.

Facilities and Equipment

We are not aware of any acute capacity issues in this area. There are some capacity challenges for larger UAS, as well as UAS radio frequency testing. Further, maritime platform maintenance infrastructure is likely insufficient for scaling USV and UUV long-range operations. Without a more focused analysis of relevant facilities, it is difficult to draw many conclusions.

Availability of Alternatives

This is an area of some high risk. We predominantly see issues with second- and third-tier suppliers. The primary second-tier issue is the availability of batteries and semiconductors.

Interviewees asserted that there may be a single domestic source of maritime INSs (with high accuracy), though we could not corroborate this. Third-tier availability issues were noted for non-Chinese lithium, carbon fiber, and titanium (for UUVs). We also were told that large UUV forging capacity was limited. We were unable to corroborate these assertions with other evidence.

When we looked at the AUVSI mission data we see a limited number of USVs that are CBRN mission capable, a limited number of UGSs, USVs, UUVs that are electronic warfare mission capable, and a limited number of USVs and UUVs that are logistics mission capable, indicating potential gaps in these areas.

U.S. Defense Industrial Base Capacity Fragility

Financial Outlook

This is an area of high risk for maritime platforms and medium risk for other platforms. Significant uncertainty and changing need signals from DoD, particularly for larger USVs and

UUVs, have potentially reduced capacity in the maritime sector of the DIB. We were not able to corroborate or quantify uncertainty in the demand signal, and this is a potential area for future research. Furthermore, our review of FactSet data against first- and second-tier suppliers (from the U.S. sample set) was not useful for this category.

Most of the first-tier suppliers in our sample set were large, financially healthy companies. Only one of 13 unique defense prime contractors was rated at medium-high risk for financial stability. Commercial primes were more likely to be small and/or privately held, but not at higher financial risk. Among the second-tier suppliers we were able to identify, six of 23 (26 percent) were medium-high risk for financial stability. Across the first- and second-tier suppliers we examined, the shares with medium-high or high credit delinquency risks were similar, ranging between 27 and 30 percent.

DoD Sales

This is a difficult category to rate due to a lack of data. We observed lower risk and more diverse sales for smaller platforms, and interviewees suggested that larger platforms tended to be more dependent on DoD sales. Yet larger platforms would tend to be produced by bigger, financially healthier companies; therefore, it is probably a medium-risk area at most. This is an area for further research.

Number of Firms in the Sector

Overall, we see lower risk to large and small UAS. We see high risk to larger UGSs and all sizes of USVs and lower risk for UUVs (if only based on the number of manufacturers per demand signal units).

Foreign Dependence

We see foreign dependence in a variety of areas: UAS motors; electro-optical sensor lenses and charge-coupled devices; LiDAR sensors; batteries; semiconductors; INSs; winches; and raw materials such as ferrite, metals, fiberglass, and some plastics (see Table 7.1 for more details). We were not able to corroborate this assessment with other evidence beyond input from our interviewees. Batteries are perhaps the most significant dependence, but we also see more niche elements like UAS motors and electro-optical sensor components.

Factor Summary

When we look across the FaC factors for the capacity assessment, we see a mix of results that are dependent on the type and size of the platform and specific to selected components and materials.

We summarize risks related to the factors in the following tables. Table 7.1 covers criticality elements.

Table 7.1. Summary of U.S. Defense Industrial Base Capacity Criticality in Terms of Risk

Factor	High Risk (Interviews and/or Analysis Suggest Significant Challenges Affecting the UxS DIB Today)	Medium Risk (Interviews and/or Analysis Suggest Moderate Challenges Affecting the UxS DIB Today)	Low Risk (Interviews and/or Analysis Suggest Minimal Challenges Affecting the UxS DIB Today)
Defense unique		• The extra-large UAS market (> 1,320 lbs) tends to be geared more toward defense only (46%) • The large UGS (2,500–3,000 lbs) is 20% defense only	• Roughly 10% of the market is defense unique. UGSs are 9%, USVs are 7%, and UUVs are 1% • The UAS market is the most defense-unique by number of platforms, at 13%
Defense design requirements	• UGSs, USVs, and UUVs do not appear to benefit from much defense-ready COTS systems	• Efforts like the Blue UAS project are mitigating challenges and providing defense-ready COTS sUAS	
Skilled labor requirements		• Common challenges in finding and retaining STEM professionals, including AI and cybersecurity specialists, along with issues in select trades such as welding	
Facilities and equipment		• Interviews noted restrictions on testing long-duration autonomous USVs and UUVs at full range	• Some capacity challenges for large UAS testing and radio frequency electronic warfare testing. • Long-term USV and UUV maintenance capabilities are likely insufficient
Materials and components with long lead times	• Smaller UxS firms are unable to compete with large buyers of semiconductors	• Long delivery times reported for specialized batteries, fiberglass, ferrite, iridium, and titanium	
Availability of alternatives	• Issues with alternatives in components, such as batteries (with third-tier suppliers of lithium), semiconductors, carbon fiber, and titanium	• There may be a single domestic source of maritime INSs (with high accuracy), though we could not corroborate this	• Limited USVs that are CBRN mission capable • Limited UGSs, USVs, and UUVs that are electronic warfare mission capable • Limited USV and UUV logistics that are mission capable

High risks to criticality are focused on material availability, component availability (semiconductors), and defense-ready COTS systems for UGSs, USVs, and UUVs. While these risks are posing significant challenges to the UxS DIB today and limit the capacity of the firms to increase deliveries, there was no indication of shortfall relative to the expected future demand signal. However, if the demand signal changes acutely (e.g., the United States moves from a peacetime to a wartime footing), then the high-risk challenges identified would likely be some of the first to present problems.

Table 7.2 summarizes fragility factors identified in this research.

Table 7.2. Summary of U.S. Defense Industrial Base Capacity Fragility in Terms of Risk

Factor	High Risk (Interviews and/or Analysis Suggest Significant Challenges Affecting the UxS DIB Today)	Medium Risk (Interviews and/or Analysis Suggest Moderate Challenges Affecting the UxS DIB Today)	Low Risk (Interviews and/or Analysis Suggest Minimal Challenges Affecting the UxS DIB Today)
Financial outlook	• Demand uncertainty is exacerbating labor challenges; early career STEM professionals leave when programs are delayed • Suppliers were more likely than prime contractors to be rated at medium-high risk for financial stability (26% and 5%, respectively) • Both suppliers and primes had a similar rate of medium-high to high risk for credit delinquency (26% and 29%, respectively)	• Interviews suggest some contraction in the USV and UUV market due to demand uncertainty • DoD ordering fewer platforms than initially anticipated leads to reduced up-front investment from firms for new acquisitions	• All prime contractors we examined were rated at low risk for financial stability except Huntington Ingalls Industries (rated at medium) based on Experian credit analysis
DoD sales		• Small volume DoD orders for sUAS are less attractive to firms, especially if they need to modify their platform (Blue UAS project is mitigating) • Some indications of USV, UUV, and large UAS DoD dependency	• No indication of a lot of dependency among UAS and UGS firms to DoD sales, but data are limited
Number of firms in the sector	• Few firms are producing large UGSs, USVs, and UUVs • Volumes in the maritime markets are simply too low to support more than a handful of vendors	• The number of firms seems limited for UGSs given the demand signal ($29 million to $83 million per firm in expected 2026 market) • The USV market is limited, but with low demand ($6 million to $50 million per firm in 2026) • The UUV market seems healthy relative to the demand signal ($6 million to $100 million per firm in 2026)	• A lot of firms for large and small UAS ($20 million to $75 million per firm in 2026)

Foreign dependence	• Batteries are a significant area of foreign dependence; U.S. integrators exist but they seem highly dependent on Chinese mineral processing and other Asian countries for lithium-ion cells (mineral deposits of lithium salt brines exist in Argentina, Australia, Chile, and China)
	• Chinese acquisitions (e.g., of the Italian UAS manufacturer in 2021) can create instant dependencies
		• Preferred foreign sources (likely short of true dependence) for things like
		  ○ UAS motors (Europe, Taiwan)
		  ○ electro-optical components (Japan)
		  ○ semiconductors (South Korea, Taiwan)
		  ○ LiDAR sensors (Switzerland)
		  ○ Ferrite, titanium, fiberglass, and some plastics	

We identify more high-risk challenges related to the fragility of the UxS DIB than to criticality, and therefore conclude that the UxS DIB is more fragile than it is critical—that is, there is more risk to losing capabilities than there is in replacing capabilities once they are gone. We note high risks to firms' financial outlook (particularly second-tier suppliers); a small number of firms producing large UGSs, USVs, and UUVs; and foreign dependence on lithium and lithium-ion batteries.

Yet despite these high risks, similar to criticality, we again do not see any indication that the risks are affecting firms' ability to meet the expected future demand signal. There are few firms producing large platforms, but there is minimal demand for such platforms. There is foreign dependence on Lithium, and it is subject to supply chain challenges (similar to other materials and components), but it is not otherwise affecting capacity relative to expected demand at the time of this writing. Therefore, we conclude that none of these fragility risks seem sufficient to be very concerned about the capacity of the DIB to produce and sustain anticipated DoD UxS demand levels. However, as we noted for criticality, with unexpected future demand shocks, such as a wartime footing, these risks may be some of the first to present problems for DoD and the UxS DIB.

Overall we see that the most important capacity challenges for OUSD(R&E) to work to address involve defense design requirements; materials and components with long lead times; availability of alternatives; financial outlook; the number of firms in the UGS, USV, and UUV sectors; and foreign dependence, particularly on batteries.

A Near-Peer Comparison

We compared the United States with China and Russia for the categories where possible. In Figure 7.1 we show the demand signal for UAS.

Figure 7.1. Comparison of Historical and Expected Military UAS Units Delivered for the United States, Russia, and China

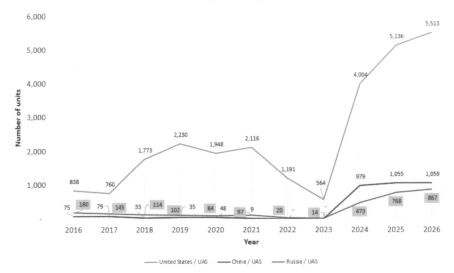

SOURCE: Authors' analysis of data from Janes Markets Forecast.

Figure 7.2 shows demand for UGS.

Figure 7.2. Comparison of Historical and Expected Military UGS Units Delivered for the United States, Russia, and China

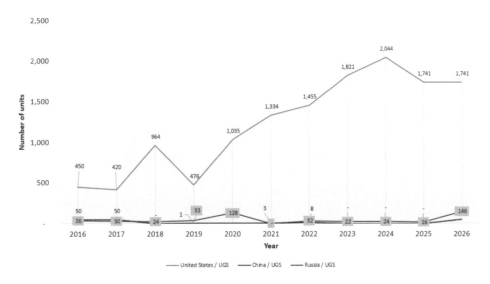

SOURCE: Authors' analysis of data from Janes Markets Forecast.

Figure 7.3 shows demand for USVs and UUVs.

Figure 7.3. Comparison of Historical and Expected USV and UUV Units Delivered for the United States, Russia, and China

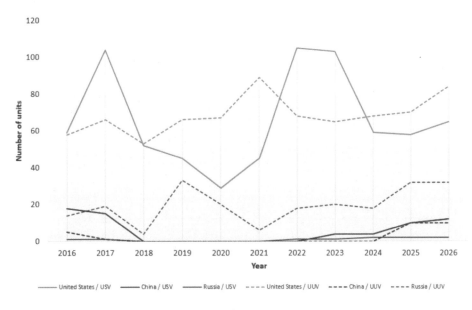

SOURCE: Authors' analysis of data from Janes Markets Forecast.

Figure 7.4 shows defense uniqueness by country. FaC typically considers defense uniqueness a driver of criticality. Firms with larger markets and diverse customer bases tend to have higher total capacity and greater financial health.

Figure 7.4. Comparison of Defense Uniqueness

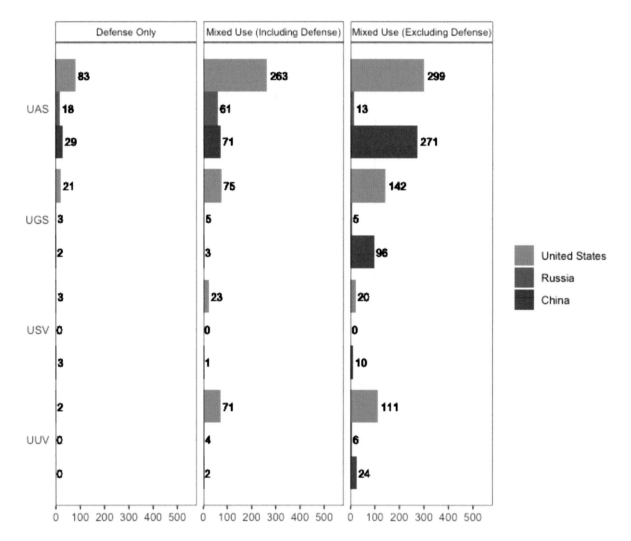

SOURCE: Authors' analysis of data from AUVSI.

We see that the China has the least defense uniqueness in UAS (per platform), whereas Russia is the most unique, with the United States in the middle. The United States and China also have relatively large proportions of nondefense UGS and UUV platforms. This implies that China may have a more resilient industrial base to potential shocks.

We also looked at the availability of alternatives between the nations, as shown in Figure 7.5.

Figure 7.5. Comparison of the Availability of Alternatives, by Mission Type

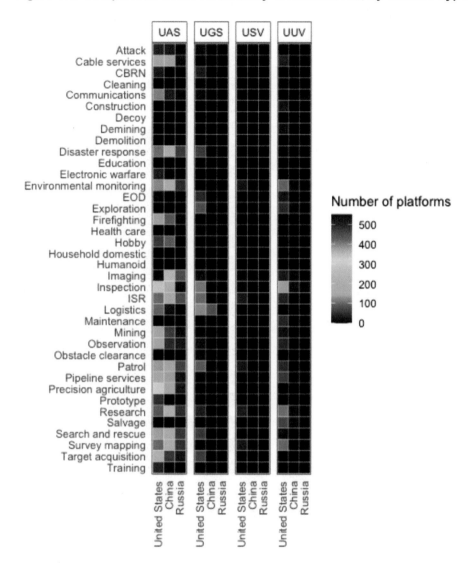

SOURCE: Authors' analysis of data from AUVSI.

There are relatively few platforms for some defense-unique missions (such as attack, CBRN, demining, and electronic warfare) across all three countries, but larger numbers for dual uses that include defense (such as imaging, ISR, and patrol).

Finally, we looked at the number of alternative producers by platform type in each country, as shown in Figure 7.6.

The number of producers of UGSs, USVs, and UUVs that are defense unique or of mixed use that include defense is very small in China and Russia. There are relatively more UAS producers, particularly in China. In comparison with Figure 7.4, it is interesting to note that in most sectors, there are only one or two unique platforms per producer—with a few exceptions, mostly among Chinese nondefense producers (3.8 platforms per UAS producer, 5.6 platforms per UGS producer, and ten platforms for one USV producer).

Figure 7.6. Comparison of the Number of Firms in the Sector

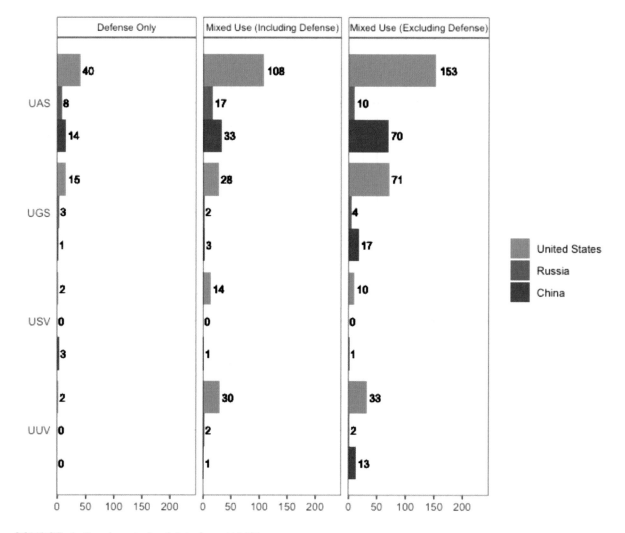

SOURCE: Authors' analysis of data from AUVSI.

Recommendations

Our recommendations are as follows:

- **Improve demand certainty for USVs and UUVs to mitigate the fragility of the maritime DIB.** There are indications that demand uncertainty has caused some contraction in the maritime market, though the number of firms aligns with the current low level of demand. The maritime market remains fragile to a demand shock.
- **Explore resilience mitigations for: lithium-ion batteries, UAS motors, electro-optical components, semiconductors, LiDAR sensors, ferrite, iridium, titanium, and fiberglass.** A combination of long lead times, limited availability of alternatives, and dependence foreign suppliers for these components and materials have been reported by vendors and program offices. Caution is warranted because vendors indicated two common themes for global sourcing: low prices and higher quality. Moving too quickly with resilience measures (e.g., insourcing, expansion of covered countries in Section 848

of the NDAA) may have unintended consequences that may include increasing cost and reduced quality.

- **Expand the DIU's Blue UAS project (or a similar effort) to include UGSs, USVs, and UUVs.** This will create more defense-ready COTS systems and better enable smaller performers to engage DoD. Bridging the gap between COTS systems and what we term here as "defense-ready" COTS systems is an important element in mitigating the fragility and criticality of capabilities. Having defense-ready capabilities mitigates fragility by increasing the number of firms in the defense sector and mitigates criticality by providing more platforms that can meet defense design requirements.
- **Explore options to boost the large UAS commercial industrial base to mitigate defense uniqueness in large UAS platforms (hurdles may exist due to FAA policies).** The number of large commercial platforms and producing firms remains low, to include large UAS. If DoD wishes to have more large defense-ready UAS, then the industrial base needs support.
- **Support policies that grow important trade skills such as electronics soldering (for UAS), and welding, forging, and metal casting (for USVs and UUVs).** These skills were identified to us as areas that are in demand and where the lack of available talent is particularly acute. Additionally, forging and casting are areas suffering from capacity challenges.

Next Steps

We propose that OUSD(R&E) consider several potential next steps:

- There may be an opportunity to develop and leverage quantitative approaches to the assessment of the skilled labor criticality factor.
- Quantifying the level of defense design specificity was a challenge beyond the scope of this project, but this is something for OUSD(R&E) to consider in the future.
- Information about component suppliers is difficult to find in public sources; therefore, additional interviews with government and industry SMEs and perhaps even a survey would be needed to identify key component suppliers across a range of UxS platforms.
- We could not assess potential limiting factors to DIB capacity expansion, such as skilled labor, facilities, and equipment. These categories were beyond the scope of our already time-intensive interviews, but this is an area of potential future study.
- Consider exploring other FaC areas, such as public finance, rate of technological change, and production expansion capacity.
- There is potential further analysis that can be done regarding the quantification of demand certainty based on historic planned procurement levels.
- Monitor Russia's import substitutions in response to sanctions as a result of the conflict in Ukraine.
- Future analyses may expand the scope of our sample sets to include additional use cases.
- Finally, further study into potential gaps in RDT&E funding for future DoD UxS platforms as firms are indicating that a lot of up-front investment on their part is required for uncertain future demand. This model may exclude smaller, less well-resourced vendors.

Appendix A. Additional Data on Exports and Estimated Future Production of Autonomous Uncrewed Systems by the United States, China, and Russia

In this appendix we provide some additional details on the manufacturers, system names, and end user countries for the top five exporters of defense UAS (in Table A.1), USVs and UUVs (in Table A.2), and UGSs (in Table A.3). Tables A.4, A.5, and A.6 show estimated production of UAS, UGSs, USVs, and UUVs by platform name in the United States, China, and Russia, respectively.

Table A.1. Top Five Exporters of Defense UAS

Country/Supplier/Model	Total Exports, 2009–2020			End User Countries
	Cost ($ Millions)	Units	Percentage of Defense Sales	
United States	**4,579.47**	**2,310**	**8%**	
AeroVironment, Inc.	**288.88**	**1,421**	**29%**	
Arcturus T-20	6.31	15		Mexico
DRAACO	9.92	55		France
Puma AE	82.27	354		Australia, Canada, Denmark, Egypt, Germany, Netherlands, Sweden
Raven	151.32	717		Belgium, Canada, Colombia, Czech Republic, Denmark, Italy, NATO, Netherlands, Norway, Philippines, Portugal, Saudi Arabia, Spain, Ukraine
Wasp	4.16	64		Spain, Sweden
Wasp AE	34.90	216		Australia, Netherlands
General Atomics	**1,083.02**	**42**	**5%**	
Predator	116.38	14		Morocco, United Arab Emirates
Predator A	43.80	4		Italy
Reaper	913.34	24		France, Italy, Spain, United Kingdom
SkyGuardian	9.50	0		India
Kratos Defense & Security Solutions	**34.05**	**44**	**8%**	
BQM-167 Skeeter	16.53	26		Singapore, Sweden

| Country/Supplier/Model | Total Exports, 2009–2020 | | | |
	Cost ($ Millions)	Units	Percentage of Defense Sales	End User Countries
MQM-107	5.01	0		South Korea, Turkey, United Arab Emirates
MQM-178 Firejet	4.91	18		Sweden, United Kingdom
XQ-58A Valkyrie	7.60	0		Unknown
Lockheed Martin Corporation	**31.38**	**113**	**2%**	
DesertHawk	7.03	100		United Kingdom
Indago	0.60	6		Switzerland
PTDS Aerostat	23.75	6		Saudi Arabia
Tethered Aerostat Radar System (TARS)	0.00	1		Philippines
Neural Robotics, Inc.	**19.73**	**48**	**100%**	
AutoCopter	19.73	48		Various
Northrop Grumman Corporation	**1,327.63**	**37**	**6%**	
BQM-34	1.23	0		Israel, Italy
BQM-74/MQM-74	6.33	0		Finland, Norway, Saudi Arabia, South Korea, Taiwan
Global Hawk	1,022.14	8		Australia, NATO, South Korea
Global Hawk Ops	260.00	0		NATO
Global Hawk Support Services	29.21	0		Pakistan
MQM-33/36	0.88	0		Finland, India
Scarab Model 324	7.83	29		Egypt
Prioria Robotics, Inc.	**7.24**	**35**	**74%**	
Maveric	7.24	35		Canada
			0%	
Raytheon Technologies Corporation	**0.40**	**3**		
Silver Fox	0.40	3		Colombia
TCOM LP	**501.90**	**73**	**71%**	
Aerostat	8.10	2		Malaysia, Singapore
TCOM LP 71M	24.36	6		India
TCOM LP Aerostats	469.44	65		Various

| Country/Supplier/Model | Total Exports, 2009–2020 | | | End User Countries |
	Cost ($ Millions)	Units	Percentage of Defense Sales	
Textron, Inc.	**53.75**	**52**	**17%**	
Aerosonde	17.02	0		Australia, Qatar
Shadow	536.73	52		Australia, Italy, Pakistan, Romania, South Korea, Sweden
The Boeing Company	**405.16**	**346**	**9%**	
Blackjack Export Version	100.43	55		Canada, Netherlands, Oman, Poland, United Arab Emirates
ScanEagle	304.74	291		Australia, Canada, Columbia, Czech Republic, Indonesia, Iraq, Italy, Japan, Brazil, Malaysia, Netherlands, Pakistan, Philippines, Poland, Singapore, United Kingdom, Vietnam
UAV Solutions, Inc.	**0.78**	**4**	**100%**	
Phoenix 30	0.78	4		Romania
Israel	**2,558.37**	**1,251**	**48%**	
Airbus Group SE	**80.02**	**4**	**100%**	
SIDM Eagle I (Harfang)	80.02	4		France (jointly developed with IAI, based on Heron)
BlueBird Aero Systems Ltd.	**10.89**	**127**	**100%**	
MicroB	4.41	59		Chile, NATO
SpyLite	6.47	68		Chile, unknown Africa, unknown Americas
Elbit Systems Ltd.	**421.34**	**381**	**40%**	
Hermes	175.03	18		Azerbaijan, Brazil, Chile, Mexico, Singapore, unknown Americas
Hermes 450	81.48	13		Colombia, Philippines, Thailand
Hermes 900	110.17	15		Azerbaijan, Colombia, European Union (EU), Philippines, Switzerland
Skylark	48.67	135		France, Mexico, Peru, Philippines, Poland, South Korea

| Country/Supplier/Model | Total Exports, 2009–2020 | | | |
	Cost ($ Millions)	Units	Percentage of Defense Sales	End User Countries
Innocon Ltd.	**33.49**	**127**	**100%**	
MicroFalcon I	20.50	102		Peru, unknown Asia
MiniFalcon I and II	12.99	20		Unknown
Israel Aerospace Industries Ltd.	**1,738.18**	**420**	**74%**	
Birdy	73.56	290		Russia
Heron	1,426.78	83		Australia, Azerbaijan, Canada, EU, Germany, Greece, India, Singapore, South Korea, Turkey, Vietnam
Heron- 1	6.24	1		Brazil
Heron TP	101.75	10		India
Searcher	129.85	36		Azerbaijan, India, Indonesia, Russia, South Korea, Spain
Rafael Advanced Defense Systems Ltd.	**265.73**	**193**	**95%**	
Aerostar	177.43	63		Mexico, Netherlands, Russia, Thailand, unknown, unknown Africa
ArrowLite	5.27	40		United States
Dominator	5.42	3		Thailand
Dominator 42	63.66	26		Mexico, unknown non–United States
Miskam	1.92	1		Canada
Orbiter	12.03	60		Peru, Poland, Russia, Vietnam
Top I Vision	**8.72**	**4**	**100%**	
Casper 200/250	1.36	0		Various
Rufus	7.36	4		Various
Italy	**743.94**	**130**	**61%**	
Leonardo SpA	**743.94**	**130**	**76%**	
Falco	136.48	38		Pakistan, Saudi Arabia, unknown Middle East
Mirach	46.99	92		Algeria, Belgium, Netherlands, United Kingdom
Project Zero	30.00	0		Unknown

| Country/Supplier/Model | Total Exports, 2009–2020 | | | |
	Cost ($ Millions)	Units	Percentage of Defense Sales	End User Countries
Rotary Wing Unmanned Air System (RWUAS) Phase 2	10.00	0		United Kingdom
Sky-Y	520.48	0		EU
France	**504.24**	**620**	**44%**	
AeroNautic Services and Engineering	**3.86**	**1**	**100%**	
T-C350	3.86	1		Unknown
Airbus Group SE	**3.22**	**18**	**3%**	
Fox TS1	0.90	0		Romania
Tracker	2.32	18		Austria
		0		
Dassault Group	**360.00**		**78%**	
Neuron	360.00	0		EU
Parrot	**0.05**	**48**	**100%**	
ANAFI UAV	0.05	48		Switzerland
Safran Group	**137.11**	**0**	**44%**	
Sperwer	137.11	0		Greece, Netherlands
China	**194.52**	**496**	**9%**	
AVIC	**85.10**	**43**	**6%**	
Wing Loong	75.64	37		Egypt, Kazakhstan, Nigeria, Saudi Arabia, United Arab Emirates
Wing-Loong I	9.46	6		Egypt, Indonesia
China Aerospace Science and Technology Corporation	**107.24**	**36**	**32%**	
CH-3	40.51	13		Algeria, Nigeria
CH-4	66.73	23		Algeria, Indonesia, Iraq, Nigeria, Saudi Arabia

	Total Exports, 2009–2020			
Country/Supplier/Model	Cost ($ Millions)	Units	Percentage of Defense Sales	End User Countries
DJI	**2.17**	**412**	**100%**	
Matrice	0.50	50		Israel
Mavic	1.06	230		Germany, Israel
Phantom 4	0.61	132		Australia, Germany

SOURCE: Janes Markets Forecast.
NOTE: A quantity of zero in the Units column indicates a nonproduction contract (e.g., R&D, services, or spares) or that the quantity is unknown.

Table A.2. Top Five Exporters of Defense USVs and UUVs

	Total Exports, 2009–2020			
Country/Supplier/Model	Cost ($ Million)	Units	Percentage of Defense Sales	End User Countries
Sweden	**257.97**	**55**	**59%**	
Air & Sea ACC Group AB	**4.48**	**3**	**58%**	
V8 ROV	4.48	3		Indonesia, South Korea
Saab AB	**253.49**	**52**	**62%**	
AUV 62F	47.77	17		Unknown
AUV62	57.60	12		Unknown
Double Eagle	107.72	14		Australia, Belgium, Denmark, Finland, France, Netherlands, Poland
Falcon	12.07	6		Colombia, Finland, Peru, Poland, South Korea
MMCM ROV (France) R&D	8.99	0		France (Maritime Mine Counter Measures)
MMCM ROV (UK) R&D	9.27	0		United Kingdom
SAM-3	10.06	3		Finland
France	**199.78**	**43**	**48%**	
Groupe Gorgé	**196.56**	**40**	**56%**	
Alister	27.40	9		EU, Kazakhstan, unknown
ECA UUV	1.01	1		Romania
H800	0.95	1		Indonesia

Country/Supplier/Model	Cost ($ Million)	Units	Percentage of Defense Sales	End User Countries
Inspector Mk.2	22.52	6		Russia, unknown Asia
K-Ster I	2.67	11		Kazakhstan, Singapore, unknown Asia
MMCM AUV (UK) R&D	36.92	0		United Kingdom
Olister	21.96	4		Malaysia
PAP	50.58	2		Greece, Japan, Pakistan, Singapore, South Africa, Turkey, United Kingdom
SeaScan	11.51	6		Russia
SWARMs project	21.03	0		EU
Subsea Tech	**2.94**	**3**	**100%**	
Catarob T-02	2.94	3		United Arab Emirates
Thales Group	**0.28**	**0**	**1%**	
MUSCLE	0.28	0		NATO
United States	**161.63**	**130**	**3%**	
Deep Ocean Engineering, Inc.	**4.30**	**2**	**100%**	
Phantom	4.30	2		Canada, Sweden
General Dynamics	**4.32**	**8**	**1%**	
Bluefin-Spray Glider	1.73	6		NATO
Training UUV	2.59	2		Finland
Huntington Ingalls Industries (Hydroid)	**65.93**	**27**	**22%**	
REMUS	49.36	18		Australia, Canada, Finland, Germany, Italy, Japan, NATO, Norway, Poland, United Kingdom, unknown
REMUS 100	17.57	9		Belgium, Japan, Netherlands, New Zealand, Singapore, Sweden

| Country/Supplier/Model | Total Exports, 2009–2020 | | | End User Countries |
	Cost ($ Million)	Units	Percentage of Defense Sales	
L3Harris Technologies, Inc.	**5.25**	**4**	**3%**	
Iver3	5.25	4		Canada
Marine Advanced Robotics, Inc.	**1.76**	**1**	**100%**	
WAM-V	1.76	1		Australia
Oceaneering International, Inc.	**3.95**	**2**		
Magnum	3.95	2		Norway
Raytheon Technologies Corporation	**17.61**	**2**	**76%**	
AN/SLQ-48	17.61	2		Greece, Taiwan
SeaRobotics Corporation	**0.50**	**1**	**28%**	
USV-2600	0.50	1		Canada
TechnipFMC plc	**6.07**	**3**	**100%**	
HD ROV	6.07	3		South Korea
Teledyne Technologies, Inc.	**8.67**	**30**	**4%**	
LBV150SE	1.68	8		Sweden
Slocum Glider	2.43	10		Australia, NATO
vLBV300	0.38	1		United Arab Emirates
vLBV950	3.47	8		Canada
Z-Boat	0.72	3		Australia
The Boeing Company (Liquid Robotics)	**7.66**	**27**	**3%**	
Wave Glider	7.66	27		Japan, NATO
The Columbia Group, Inc.	**11.50**	**0**	**100%**	
Pluto	11.50	0		Egypt
VideoRay	**2.25**	**17**	**16%**	
MSS Defender	0.90	8		India

Country/Supplier/Model	Total Exports, 2009–2020			End User Countries
	Cost ($ Million)	Units	Percentage of Defense Sales	
VideoRay Pro 4	1.09	9		Italy, Netherlands, United Kingdom
VideoRay Pro II	0.25	0		Taiwan
Zyvex Technologies	**20.87**	**5**	**100%**	
Piranha	20.87	5		Singapore
Norway	**80.29**	**17**	**47%**	
Argus Remote Systems AS	**1.61**	**0**	**100%**	
Bathysaurus	1.16	0		Sweden
Rover Mk II	0.45	0		Sweden
Egersund Group AS	**1.98**	**1**	**100%**	
SUB-fighter ROV	1.98	1		Brazil
Kongsberg Gruppen ASA	**77.70**	**14**	**50%**	
C'Inspector	2.69	2		Turkey
HUGIN	74.01	12		Finland, India, Indonesia, Italy, Peru, Poland
Iceland	**68.30**	**19**	**100%**	
Teledyne Technologies, Inc. (Hafmynd ehf)	**68.30**	**19**	**100%**	
Gavia	68.30	19		Australia, Canada, Denmark, Poland, Portugal, Russia, Turkey, United Arab Emirates, United States

SOURCE: Janes Markets Forecast.
NOTE: A quantity of zero in the Units column indicates a nonproduction contract (e.g., R&D, services, or spares) or that the quantity is unknown.

Table A.3. Top Five Exporters of Defense UGSs

Country/Supplier/Model	Total Exports, 2009–2020			End User Countries
	Cost ($ Million)	Units	Percentage of Defense Sales	
United States	**1,036.33**	**7,990**	**27%**	
Chemring Sensors & Electronic Systems	**7.46**	**12**	**100%**	
HMDS	3.81	6		Canada
Minestalker II	3.65	6		Unknown
Hippo Multipower	**0.44**	**2**	**100%**	
Hippo X Unmanned	0.44	2		United Kingdom
L3Harris Technologies	**89.68**	**122**	**100%**	
T7	89.68	122		United Kingdom
Mesa Associates, Inc.	**6.47**	**35**	**96%**	
Matilda	6.47	35		Various
Northrop Grumman Corporation	**355.38**	**1,485**	**63%**	
Andros	297.35	1,315		Unknown, various
Andros F6B	2.03	10		Canada
CaMEL	22.86	60		Israel
HD-1	0.54	3		Iraq
Mini-ANDROS II	0.16	0		Iraq
Scorpion RCV	0.08	0		Israel
V-A1	25.31	76		Unknown
Wolverine	7.05	21		South Korea, Taiwan
QinetiQ North America (Foster-Miller)	**396.56**	**1,799**	**37%**	
Dragon Runner	38.33	220		Australia, Netherlands, United Kingdom
Dragon Runner 20	1.24	4		Canada
MTRS	69.83	100		Australia
Talon	262.29	1,385		Australia, Netherlands, Poland, Turkey, various
Talon IV	24.87	110		Iraq

| Country/Supplier/Model | Total Exports, 2009–2020 | | | End User Countries |
	Cost ($ Million)	Units	Percentage of Defense Sales	
ReconRobotics	**105.14**	**4,196**	**65%**	
Recon Scout	105.14	4,196		United Kingdom, unknown
Safariland LLC	**10.66**	**108**	**96%**	
Digital Vanguard	9.40	108		Various
Vanguard	1.26	0		Brazil, Canada, Colombia, Singapore, unknown Europe, various
Teledyne FLIR LLC	**64.53**	**231**	**6%**	
FirstLook	2.45	48		Germany
Kobra	15.68	15		Unknown
PackBot	42.20	143		Brazil, France, Iraq, Norway, Spain, United Kingdom
PackBot 510	4.20	25		Australia, Canada
Germany	**359.33**	**592**	**92%**	
Unmanned Systems Investments GmbH	**359.15**	**592**	**100%**	
Telemax	8.91	24		Australia, Israel, United States
Telemax Hybrid	0.38	2		Bangladesh
tEODor EVO	9.80	21		Norway
tEODor Robot	340.06	545		Argentina, Australia, Austria, Bahrain, Belgium, Canada, Czech Republic, Denmark, Finland, France, Greece, Indonesia, Kazakhstan, Kuwait, Nigeria, Norway, Oman, Poland, Portugal, Qatar, Romania, Saudi Arabia, Spain, Sweden, Switzerland, United Arab Emirates, unknown
KAPPA Opto-electronics	**0.18**	**0**	**100%**	
SON	0.18	0		Russia
Croatia	**87.82**	**123**	**100%**	
DOK-ING Ltd.	**87.82**	**123**	**100%**	
M160	78.17	109		United States
MV Mini-Flail	0.19	0		Iraq, Sweden
MV-10	5.66	8		Australia
MV-4	3.81	6		Colombia, South Korea

| Country/Supplier/Model | Total Exports, 2009–2020 | | | End User Countries |
	Cost ($ Million)	Units	Percentage of Defense Sales	
United Kingdom	**79.41**	**134**	**39%**	
Aardvark Clear Mine Ltd.	**7.12**	**20**	**100%**	
Aardvark	7.12	20		Unknown
ABP	**6.91**	**44**	**100%**	
Guardian	6.55	43		Various
Sentinel	0.35	1		China
Armtrac Limited	**11.29**	**14**	**100%**	
Armtrac	2.62	4		Saudi Arabia
Armtrac 400	8.67	10		Egypt, Nigeria
Northrop Grumman Corporation (Alvis Logistics)	**54.06**	**56**	**100%**	
Wheelbarrow	54.06	56		Bahrain, India, Iraq, Kuwait, New Zealand, Oman, Portugal, Saudi Arabia, Singapore, United Arab Emirates, unknown
QinetiQ Group Plc	**0.03**	**0**		
Carver	0.03	0		United States
Israel	**60.05**	**552**	**16%**	
Automotive Robotic Industry Ltd.	**18.27**	**45**	**48%**	
Amstaf	18.27	45		India, South Korea
General Robotics Ltd.	**0.89**	**80**	**64%**	
DOGO	0.89	80		France, India
Roboteam	**40.89**	**427**	**76%**	
MTGR	38.12	415		Japan, Poland, Singapore, Thailand, United Kingdom, United States
Probot	1.10	5		France
TIGR	0.79	3		Italy
Torch-Powered Probot	0.88	4		United Kingdom

SOURCE: Janes Markets Forecast.
NOTE: A quantity of zero in the Units column indicates a nonproduction contract (e.g., R&D, services, or spares) or that the quantity is unknown.

Table A.4. U.S. Specific Platform Names and Estimated Production, 2021–2026

Platform Name	Platform Type	Estimated Production
SRM-UAS	UAS	3,000
SBS micro-UAS—production	UAS	2,700
Black Hornet—follow-on	UAS	2,000
PD-100 Black Hornet	UAS	1,805
MRM UAS	UAS	1,750
BroadSword	UAS	1,576
Army Aerial Targets	UAS	760
Air Force Next Gen sUAS	UAS	650
VTOL SURSS	UAS	600
CICADA	UAS	550
RQ-11B	UAS	410
Micro-UAVs—USMC	UAS	400
BQM-177 Sub-sonic aerial target (SSAT)	UAS	330
Advanced Subscale Aerial Platform	UAS	300
Low Cost Attritable Strike UAS	UAS	300
Runway-independent tactical UAS	UAS	255
BQM-167 Skeeter	UAS	210
LRRS UAS	UAS	180
Phantom 4	UAS	150
SilentEyes	UAS	150
MQ-9 Reaper	UAS	148
Group 1 Medium Range/Medium Endurance (MR/ME)	UAS	124
Triton	UAS	121
Banshee Jet – 80	UAS	120
Advanced UAS	UAS	103
Puma All Environment (AE)	UAS	85
QF-16	UAS	83
Group 1 VTOL Short Range/ Short Endurance (SR/SE)	UAS	71
MQ-21A Blackjack (attrition)	UAS	65
MQ-25A Stingray	UAS	65
Magnetic anomaly detector (MAD) UAV	UAS	60
RQ-7B V2 with Block III engine	UAS	36
MQ-8C	UAS	33
Long Range/Long Endurance (LR/LE)	UAS	32
Shadow M2	UAS	32

Platform Name	Platform Type	Estimated Production
Aerial targets—U.S. Navy	UAS	30
Spain Mini UAV	UAS	27
Blackwing	UAS	24
Vanilla 001	UAS	24
MQ-1C Gray Eagle	UAS	21
Army Gray Eagle Replacement	UAS	20
Group 1 Micro VTOL	UAS	20
Group Nano VTOL	UAS	17
XQ-58A Valkyrie	UAS	15
R80D SkyRaider	UAS	12
USMC Delivery Drone	UAS	10
NG RPV	UAS	9
MQ-1 Predator A, MQ-9 Reaper	UAS	8
DHS Aerostat	UAS	6
Blue Water Maritime Logistics UAS	UAS	5
Next-Gen MALE UAV	UAS	3
MQ-1/9	UAS	1
MQ-8	UAS	1
V-247	UAS	1
Standoff Robotic Explosive Hazard Detection System (SREHD)	UGS	3,190
LR Squared (LR2)	UGS	1,460
MTRS Increment II—Centaur	UGS	1,453
SO USMC UGV	UGS	1,060
CRS(I)-SPUR	UGS	1,032
SMET UGV Production Phase	UGS	624
Kobra	UGS	336
MTRS Recap	UGS	301
SO Robotic Combat Vehicle (RCV) Heavy	UGS	202
SO RCV L	UGS	202
Micro tactical ground robot (MTGR)	UGS	75
AEODRS—Increment 2	UGS	70
SUGV 310 recap	UGS	45
110 FirstLook	UGS	36
AEODRS—Increment 3	UGS	25
Ground/Transportable	UGS	24
Dragon Runner 10 recap	UGS	19

Platform Name	Platform Type	Estimated Production
SO Large EOD Robot	UGS	16
M8 8x8 Autonomous Electrical Amphibious Vehicle	UGS	5
RCV L	UGS	2
UVRS	UGS	2
Ripsaw M-5E	UGS	1
SO Engineering UGV	UGS	1
High Speed Maneuverable Surface Target (HSMST)	USV	297
Common Unmanned Surface Vehicle (CUSV) (MCM USV)—FRP	USV	60
CUSV (MCM USV)	USV	8
LRUSV	USV	6
AN/SLW-32	USV	5
LWUUVs for Multi-Function MCM USV	UUV	72
LBS-Gliders (Improved version)	UUV	60
LBS-Gliders (Replacement units)	UUV	36
Teledyne AUV	UUV	32
REMUS-Swordfish U.S. Navy	UUV	27
Long range glider	UUV	22
Knifefish FRP	UUV	21
REMUS-Kingfish U.S. Navy	UUV	21
Black Pearl UUV	UUV	20
vLBV300	UUV	20
LBS-Gliders	UUV	15
MSS Defender	UUV	15
Flying Sea Glider	UUV	12
Maritime Expeditionary Mine Countermeasures Unmanned Undersea Vehicle (MEMUUV)	UUV	12
Shallow Water Submarine Simulator	UUV	12
TETRA remote operated vehicle (ROV)	UUV	11
Knifefish LRIP	UUV	10
LBS-AUV(S)/Razorback (torpedo tube launch and recover variant)	UUV	6
Wave Glider SV3	UUV	6
XLUUV Phase II	UUV	5
SUUV	UUV	4
XLUUV Phase III	UUV	4
MEDUSA	UUV	3
REMUS-Swordfish	UUV	3

Platform Name	Platform Type	Estimated Production
LBS-AUV(S)/Razorback (dry dock shelter variant)	UUV	2
Outland ROV-2500	UUV	2
REMUS 300 (SUUV)	UUV	2
Phantom T5 Super Vectored ROV	UUV	1

SOURCE: Janes Markets Forecast.
NOTE: Platform names are as they appear in Janes.

Table A.5. Chinese Specific Platform Name and Estimated Production, 2021–2026

Platform Name	Platform Type	Estimated Production
China Army Micro UAV	UAS	4,300
China Army Close In UAV	UAS	1,710
China Army VTOL UAV	UAS	417
China Army TUAV	UAS	308
China Air Force MALE UAV	UAS	285
Aerostat	UAS	88
Long Endurance Stealth UAV	UAS	61
China UCAV	UAS	50
China Navy VTOL UAV	UAS	42
ASN-7	UAS	30
Wing-Loong I	UAS	30
China Troop Supply VTOL UAV Follow-on	UAS	30
Dark Sword	UAS	26
Long Haul Eagle	UAS	18
AT200	UAS	10
Blue Fox	UAS	10
AV500W	UAS	10
X-Chimera	UAS	8
CH-92	UAS	6
TB001	UAS	6
Zhanfu H16-V12	UAS	6
ZC300	UAS	5
LJ-1	UAS	2
Caihong Solar UAV	UAS	1
Large UGV	UGS	50
Sharp Claw I	UGS	6
SO Q-UGV	UGS	5

117

Platform Name	Platform Type	Estimated Production
Marine Lizard	USV	36
USV (China)	USV	16
Type 529 USV	USV	12
HSU001	UUV	16
ROV (Type 082II MCMVs)	UUV	4

SOURCE: Janes Markets Forecast.

Table A.6. Russian Specific Platform Name and Estimated Production, 2021–2026

Platform Name	Platform Type	Estimated Production
Russia Army Micro UAV	UAS	1,100
Russia Army Small UAV	UAS	550
Russia Army UAV	UAS	168
Russia SR VTOL UAV	UAS	60
Small Tactical VTOL UAV	UAS	60
Forpost	UAS	57
Orlan-30	UAS	50
Russia Armed MALE Prod	UAS	36
Loyal Wingman—Russia	UAS	30
Russia MR VTOL UAV	UAS	18
Russia Navy VTOL UAV	UAS	16
Russia Large VTOL UAV	UAS	15
Orion-E	UAS	12
Russia HALE	UAS	12
BPV-500	UAS	12
Granat-4	UAS	10
Orlan-10	UAS	10
The KBLA-IVT	UAS	8
Eleron-3SV	UAS	7
Russia UCAV Production	UAS	5
S-70 Okhotnik	UAS	5
Russia Aerostat	UAS	3
Medium UGV	UGS	75
Large mine clearing UGV	UGS	50
Light UGV	UGS	50
Uran-9	UGS	22
SO Russian Federation CBRN UGV	UGS	15

Platform Name	Platform Type	Estimated Production
SO EOD UGV	UGS	14
Large UGV	UGS	5
SO Russian Federation Marker Combat UGV	UGS	5
SO Russian Federation Soratnik Combat UGV	UGS	5
SO Large mine clearing/engineering UGV	UGS	1
SO Medium engineering UGV	UGS	1
USV (Minesweeper Vessel)	USV	8
UUV (Russia)	UUV	22
UUV (Mine disposal)	UUV	20
Concept-M AUV	UUV	8
AUV (Project 20386)	UUV	5
ISPUM Unmanned MCM	UUV	2
Klavesin-2P-PM	UUV	2
ROV (Project 21301)	UUV	2
ROV (Yantar class vessel)	UUV	2
UUV/ROV (Project 02670 class)	UUV	1

SOURCE: Janes Markets Forecast.

Appendix B. Size Categories and Platform Sample Sets

In this appendix we detail the sample sets of platforms used to help scope the project. We have four sets: United States, China, Russia, and international. We developed these size categories using U.S. government size categories for each UxS domain, as shown in Table B.1.

Table B.1. Size Categories of UAS Platforms

Class	Maximum Gross Takeoff Weight	Operating Altitude	Example Crafts
Extra-large (Groups 4 and 5)	>1,320 lbs	> FL180	Group 4: Fire Scout (MQ-8B, RQ-8B), Predator (MQ-1A/B) Group 5: Reaper (MQ-9A), Global Hawk (RQ-4)
Large (Group 3)	55–1,320 lbs	< FL179	Shadow (RQ-7B), Tier II/STUAS
Medium (Group 2)	21–55 lbs	< 3,500 ft AGL	ScanEagle
Small (Group 1)	0–20 lbs	< 1,200 ft AGL	Raven (RQ-11), Wasp

SOURCE: Code of Federal Regulations, Title 14, Chapter 1, Subchapter F, Part 107, Small Unmanned Systems, 2021, last amended June 2, 2022.
NOTES: FAA regulations define small UAS as those under 55 lbs. For comparative simplicity, we merged groups 4 and 5. AGL = above ground level; FL = flight level.

Table B.2. Size Categories of UGS Platforms

Class	Maximum Weight	Example Crafts
Extra-large	> 30,000 lbs	Deuce s/SRS, Type-X
Large (heavy)	2,501–30,000 lbs	ARTS, T3 Dozer w/SRS
Medium	401–2,500 lbs	RONS, Titan
Small (light)	0–400 lbs	TALON, MATILDA

SOURCES: O'Rourke, 2021; Robotic Systems Joint Project Office, *Unmanned Ground Vehicle (UGV) Interoperability Profile (IOP) Overarching Profile*, Warren, Mich., December 21, 2011; DoD, *Unmanned Systems Integrated Roadmap, 2013–2038*, Washington, D.C., August 28, 2014; DoD, *Unmanned Systems Integrated Roadmap, 2017–2042*, Washington, D.C., August 28, 2018.

Table B.3. Size Categories of USV Platforms

Class	Length	Example Crafts
Extra-large (Class 4)	> 50 m	Ghost Fleet, Overload LUSV
Large (Class 3)	> 12–50 m	Sea Hunter 2, MDUSV
Medium (Class 2)	> 7–12 m	MHU 1–4
Small (Class 1)	≤ 7 m	GARC Optionally Manned, ADARO/ MUSCL

SOURCES: DoD, 2014; DoD, 2018; UAS Task Force, Airspace Integration Integrated Product Team, *Unmanned Aircraft System Airspace Integration Plan*, version 2.0, Washington, D.C.: U.S. Department of Defense, March 2011.

Table B.4. Size Categories of UUV Platforms

Class	Diameter	Example Crafts
Extra-large	> 84 inches	Office of Naval Research Innovation Naval Prototype, Orca
Large	> 21–84 inches	SNAKEHEAD Ph1 Vehicle
Medium	> 10–21 inches	LBS-AUV, LBS-G, Knifefish
Small	> 3–10 inches	Mk18 Mod 2, IVER, Sandshark

SOURCES: DoD, 2014; DoD, 2018; UAS Task Force, 2011.

The sample-set tables have eight columns, defined as follows:

- user type (defense or commercial industrial base)
- use case (main operation mission of the UxS)
- type (UxS domain)
- size category (size of the UxS)
- future option (the anticipated future UxS platform for the mission)
- prime supplier (supplier of the future UxS platform)
- legacy option (an existing UxS that is anticipated to be phased out)
- legacy supplier (supplier of the legacy UxS platform).

Each sample set has 18 combinations of parameters pertaining to user type, use case, type, and size.

U.S. Sample Set

Table B.5 lists the UxS platforms used in the U.S. sample set. There are many potential use cases we could not consider for scoping reasons (e.g., UAS explosive attack, which would include loitering munitions). Not all legacy options were filled in, as it is unlikely that these systems will be indicative of the future U.S. demand signal.

121

Table B.5. U.S. UxS Sample Set

User Type	Use Case	Type	Size	Future Option	Prime Manufacturer	Legacy Option
Defense	Surveillance and reconnaissance	UAS	Group 3	Aersonde HQ	Textron	RQ-21 Blackjack by Boeing Insitu
Defense	Surveillance and reconnaissance	UAS	Group 5	RQ-170	Lockheed Martin	RQ-4 Global Hawk by Northrop Grumman
Commercial	Agricultural	UAS	Group 2	Quantix Mapper VTOL	Draganfly[a]	
Commercial	Agricultural	UAS	Group 3	ScanEagle	Boeing Institute	
Commercial	Municipal	UAS	Small (medium)	US-1	Impossible Aerospace	
Recreational	Photo hobbyist	UAS	Group 2	Alta X	Freefly	
Recreational	Photo hobbyist	UAS	Group 1	Skydio 2[b]	Skydio	
Defense	EOD	UGS	Small (light)	Centaur[c]	FLIR	
Defense	Ground combat	UGS	Small (medium)	S-MET/MUTT	General Dynamics Land Systems	
Commercial	Search and rescue	UGS	Small (light)	SUGV	FLIR	
Defense	Maritime	USV	Medium (Class 3)	Sea Hunter II	Vigor Industrial	
Defense	Maritime	USV	Large (Class 4)	Ghost Fleet Overlord[d]	L3 Harris and Gibbs & Cox	
Commercial	Maritime Multiuse	USV	Small (Class 1)	HYCAT	Sea Robotics	
Commercial	Sea floor mapping	USV	Large (Class 4)	Surveyor	Saildrone	
Defense	Maritime	UUV	Extra-large	Orca	Huntington Ingalls Industries	Echo Voyager by Boeing
Commercial	Sea floor mapping	UUV	Medium	Artemis	Phoenix International[e]	Dorado-class UUVs by Monterey Bay Research Institute[f]
Commercial	Sea floor mapping	UUV	Large	Black Pearl	General Dynamics	
Defense	Maritime mine countermeasures	UUV	Medium	Knifefish	General Dynamics	Bluefin 21 by Bluefin Robotics

[a] Draganfly, "Hybrid VTOL: Quantix™ Mapper," webpage, undated.
[b] Skydio, "Skydio 2," webpage, undated.
[c] Mike Ball, "Additional Centaur UGVs to Provide EOD to US Military," Unmanned Systems Technology, May 18, 2021.
[d] Justin Katz, "Navy Awards 'Overlord' Second-Phase Contract Mods to Undisclosed Companies," *Inside Defense*, October 2, 2019; U.S. Department of Defense, "Ghost Fleet Overlord Unmanned Surface Vessel Program Completes Second Autonomous Transit to the Pacific," press release, Washington, D.C., June 7, 2021.
[e] Phoenix International Holdings, Inc., "Phoenix International Holdings, Inc. (Phoenix) Has Been Contracted by the Commonwealth of Australia Under the Direction of Australian Transport Safety Bureau (ATSB) to Provide Deepwater Towed Side Scan Sonar Services," press release, Largo, Md., January 29, 2016.
[f] Monterey Bay Aquarium Research Institute, "Seafloor Mapping AUV," webpage, undated.

Some of these systems, such as the Ghost Fleet Overlord, are not currently deployed; however, their production and future intended operation determined its inclusion in the above list. The Strategic Capabilities Office recently ended the program and transferred the USVs to the Navy.[1]

China Sample Set

We conducted a similar analysis of Chinese systems. We were able to obtain detailed data on a large variety of UAS operating in both the defense and commercial spaces in China, but there was particularly little information available on UxS in the sea domain and even less indication that China is currently fielding UGVs. Therefore, several of the systems listed in Table B.3 may be in the concept development stage or operating as a single platform rather than as a group of deployed units.

Table B.6. Chinese UxS Sample Set

User Type	Use Case	Type	Size	Future Option	Prime Manufacturer	Legacy Option
Defense	Surveillance and reconnaissance	UAS	Group 3	AV500W[a]	China Helicopter Research and Development Institute	ASN-7 by Xi'an ASN Tech Group
Defense	Surveillance and reconnaissance	UAS	Group 5	WJ-700[b]	Haiying General Aviation Equipment	GJ-2 by Chengdu Aircraft Industry Group
Commercial	Agricultural	UAS	Group 3	XPlanet Agricultural UAS[c]	XAG	
Commercial	Agricultural	UAS	Group 2	AGRAS MG-15[d]	DJI	
Commercial	Municipal	UAS	Small (medium)	MMC Drone[e]	Shenzhen MicroMultiCopter	Infiltrator by Ziyang UAV Co., Ltd.
Recreational	Photo hobbyist	UAS	Group 2	Phantom 4 Pro v2.0[f]	DJI	
Recreational	Photo hobbyist	UAS	Group 1	Mavic Air 2[g]	DJI	
Defense	EOD	UGS	Small (light)	Unnamed bomb disposal robot[h]		EDO Drone by Research Institute of Engineering Corps
Defense	Ground combat	UGS	Small (medium)	Coastal defense robot[i]	PLA describes it as "built by a deputy company commander, Captain Zhou Jianlong"[j]	
User Type	Use case	Type	Size	Future Option	Prime Manufacturer	Legacy Option
Commercial	Search and rescue	UGS	Small (light)	Unable to find comparable system		
Defense	Maritime	USV	Large (Class 4)	Uncrewed Sea Hunter clone[k]	Jiang Tongfang New Shipbuilding Co., Ltd.	

---

[1] Justin Katz, "SCO Ends Project Overlord, Shifts Unmanned Vessels to Navy," *Breaking Defense*, January 12, 2022b.

User Type	Use Case	Type	Size	Future Option	Prime Manufacturer	Legacy Option
Defense	Maritime	USV	Medium (Class 3)	Look Out II[i]	Yunzhou Tech	Tianxiang One by Atmospheric Observation Technology Center of China[m]
Commercial	Maritime security	USV	Large (Class 4)	Uncrewed Sea Hunter clone[n]	Jian Tongfang New Shipbuilding Co., Ltd.	
Commercial	Maritime security	USV	Small (Class 2)	Uncrewed patrol boat[o]	Anhui CAS— Huacheng Intelligent Technology	
Defense	Maritime mine countermeasures	UUV	Medium	Unable to find comparable system		
Defense	Maritime	UUV	Extra-large	HSU001[p]		
Commercial	Sea floor mapping	UUV	Medium	Haiyi sea gliders[q]		
Commercial	Sea floor mapping	UUV	Large	100 ft extra-large UUVs[r]	CAS Shenyang Institute of Automation and Deepfar Ocean Technologies	

[a] "The UAV is a 'weaponized variant of its civilian-model AV500 with improved performance.' The AV500W can be configured to carry several air-to-ground weapons on stub wings mounted on either side of its fuselage"; Nurkin, Bedard, Clad, Scott, and Grevatt, 2018, p. 162.

[b] This is a domestic military version of the Wing Loong II; see Liu Zhen and Kristin Huang, "China's High-Speed, Armed Reconnaissance Drone Completes Maiden Flight," *South China Morning Post*, January 13, 2021.

[c] XAG, "XAG XPlanet Agricultural UAS: The Ultimate Agricultural Drone," webpage, undated.

[d] The AGRAS MG-15 is listed as being particularly good for crop dusting; see Jonathan Feist, "Best Commercial Drones—Surveying, Mapping, and Search & Rescue," DroneRush, March 23, 2022a.

[e] Chinese drones were used to monitor citizens to prevent infractions of pandemic restrictions and have aided in the arrest of drug dealers; see Zak Doffman, "This New Coronavirus Spy Drone Will Make Sure You Stay Home," *Forbes*, March 5, 2020.

[f] The Phantom 4 Pro v2.0 was rated as best large consumer drone by DroneRush in 2021; see Jonathan Feist, "DJI Phantom 4 Pro Review—A Superb Airframe," DroneRush, April 21, 2021.

[g] The Mavic Air 2 is rated among the best consumer photography drones for hobbyists; see Jonathan Feist, "Best Camera Drones: Photos from the Sky in Any Price Bracket," DroneRush, July 15, 2022b. It made the 2021 *Drone Girl* list; for the most recent list, see Sally French, "The Best Drones for Photographers of 2022," The Drone Girl, May 17, 2022b.

[h] Liu Xuanzun, "PLA Commissions New-Type Robot for Bomb Disposal," *Global Times*, March 31, 2021a.

[i] "The vehicle is equipped with cameras, runs on rubber tracks, and is armed with a customized Type 95 auto rifle as its turret"; Liu Xuanzun, "Self-Made Recon Robot Tested for Coast Patrol, Shows PLA's 'Grassroots Innovation for Modernization,'" *Global Times*, June 1, 2021b.

[j] Liu Xuanzun, 2021b.

[k] This was designed as a copy of the U.S. Sea Hunter, and it is unclear whether it is a commercial or defense system; see H. I. Sutton, "Chinese Navy Crafts Unmanned Sea Hunter Knock-Off," *USNI News*, September 25, 2020.

[l] Liu Xuanzun, "China's First Unmanned Missile Boat Revealed at Airshow China 2018," *Global Times*, November 7, 2018.

[m] This system is a meteorological and security drone used in the 2008 Beijing Olympics. Naval Drones, "Tianxiang One USV," webpage, undated.

[n] Designed as a copy to the U.S. Sea Hunter, it is unclear whether it is a commercial or defense system; see Sutton, 2020.

[o] Lin and Zhang, 2018.

[p] A newer system, announced by China in 2019. It is unknown how many are in use. David R. Strachan, "China Enters the UUV Fray," *The Diplomat*, November 22, 2019.

[q] Liu Caiyu, "Underwater Glider Haiyi 1000 Completes Mission," *Global Times*, October 15, 2017.

[r] This platform has been announced by China, but the manufacturer and specifics have not been disclosed; see Wilson, 2019.

The Chinese sample set indicates that China has many UxS platforms for a variety of uses, but it lacks an analogous system for small UGS search and rescue missions, as well as a medium UUV for maritime countermine missions. Furthermore, two of the platforms' closest analogous system is an unnamed copy of the U.S. Sea Hunter USV. Comparably, the list of commercial UAVs is dominated by Chinese systems, primarily from Chinese drone company DJI.[2]

Table B.7 lists the Chinese sample-set producers and their international suppliers that we were able to identify via FactSet. This includes parent companies and other affiliations between companies identified in Figure 4.5. Multiple entries indicate multiple orders from the same company.

Table B.7. Chinese Sample-Set Producers and Suppliers

Customer	Supplier	Supplier Country	Supplier Industry	Industry Category
AVIC	AMAG Austria Metall AG	Austria	Aluminum	Aluminum
AVIC	AMAG Austria Metall AG	Austria	Aluminum	Aluminum
State Grid Corporation of China	Investa Office Fund	Australia	Real estate investment	Real estate
State Grid Corporation of China	Bill Identity Ltd.	Australia	Real estate investment	Real estate
EHang Holdings Ltd.	FACC AG	Austria	Aircraft parts	Aviation
EHang Holdings Ltd.	FACC AG	Austria	Aircraft parts	Aviation
State Grid Corporation of China	Sigdo Koppers SA	Chile	Unknown (conglomerate)	Unknown
Huawei Technologies Co., Ltd.	Stroeer SE & Co. KgaA	Germany	Marketing	Marketing
State Grid Corporation of China	Software AG	Germany	Software	Software
CAS	GomSpace Group AB	Denmark	Nanosatellites and miniature radio technology	Satellites
AVIC	Safran SA	France	Aircraft parts	Aviation
AVIC	Figeac Aero SA	France	Aeronautical subassemblies	Aviation
Huawei Technologies Co., Ltd.	Balyo SA	France	Autonomous lift trucks (UGVs)	UGVs
AVIC	Safran SA	France	Aircraft parts	Aviation

---

[2] Sally French, "The Best Drones for Kids in 2022," The Drone Girl, January 7, 2022a; Jim Fisher, "The Best Drones for 2021," *PCMag*, last updated May 24, 2022; Feist, 2022b; Jonathan Feist, "Best Drones 2022: Affordable 5.4K Video Is a Reality!" DroneRush, July 18, 2022c.

Customer	Supplier	Supplier Country	Supplier Industry	Industry Category
AVIC	Figeac Aero SA	France	Aeronautical subassemblies	Aviation
AVIC	Zotefoams Plc	United Kingdom	Aerospace (polyolefin foams)	Aviation
AVIC	Zotefoams Plc	United Kingdom	Aerospace (polyolefin foams)	Aviation
CAS	AIXTRON SE	Germany	Semiconductor manufacturing equipment	Semiconductors
China Aerospace Science and Technology Corporation	Gilat Satellite Networks Ltd.	Israel	Satellite ground stations	Satellites
AVIC	TXT e-solutions SpA	Italy	Software development	Software
AVIC	Prima Industrie SpA	Italy	Laser and sheet metal machinery	Industrial materials
State Grid Corporation of China	Prysmian SpA	Italy	Fiber-optic cables	Fiber optics
AVIC	TXT e-solutions SpA	Italy	Software development	Software
AVIC	Prima Industrie SpA	Italy	Laser and sheet metal machinery	Industrial materials
China Electric Equipment Group Co., Ltd.	SFC Co., Ltd.	South Korea	Chemicals	Chemicals
State Grid Corporation of China	Hyosung Heavy Industries Corporation	South Korea	Electronics equipment	Electronics
State Grid Corporation of China	Hyosung Heavy Industries Corporation	South Korea	Electronics equipment	Electronics
State Grid Corporation of China	Airbus SE	Netherlands	Aircraft parts	Aviation
State Grid Corporation of China	Redes Energeticas Nacionais SGPS SA	Portugal	Energy infrastructure	Energy
China Aerospace Science and Technology Corporation	GomSpace Group AB	Denmark	Nanosatellites and miniature radio technology	Satellites
Tsinghua Tongfang Co., Ltd.	Technovator International Ltd.	Singapore	Energy infrastructure	Energy
AVIC	Himax Technologies, Inc.	Taiwan	Semiconductors	Semiconductors
Huawei Technologies Co., Ltd.	Nanoplus Ltd.	Taiwan	Nanometer materials	Industrial materials

Customer	Supplier	Supplier Country	Supplier Industry	Industry Category
AVIC	Himax Technologies, Inc.	Taiwan	Semiconductors	Semiconductors
EHang Holdings Ltd.	Vodafone Group Plc	United Kingdom	Telecommunications	Telecommunications
SZ DJI Technology Co., Ltd.	Skyworks Solutions, Inc.	United States	Semiconductors	Semiconductors
SZ DJI Technology Co., Ltd.	Ambarella, Inc.	United States	Semiconductors	Semiconductors
EHang Holdings Ltd.	Facebook, Inc.	United States	Technology	Technology
EHang Holdings Ltd.	Twitter, Inc.	United States	Technology	Technology
Ehang Holdings Ltd.	Alphabet, Inc.	United States	Technology	Technology
AVIC	AMETEK, Inc.	United States	Electronic instruments and electromechanical devices	Electronics
AVIC	Astronics Corporation	United States	Aerospace electronics	Aviation
AVIC	Hexcel Corporation	United States	Industrial materials	Industrial materials
AVIC	Aleris Corporation	United States	Aluminum	Aluminum
Xi'an Aerospace Technology Co., Ltd.	Precision Castparts Corporation	United States	Industrial goods and metal fabrication	Industrial materials
AVIC	AMETEK, Inc.	United States	Electronic instruments and electromechanical devices	Electronics
AVIC	Astronics Corporation	United States	Aerospace electronics	Aviation
AVIC	Hexcel Corporation	United States	Industrial materials	Industrial materials
AVIC	Aleris Corporation	United States	Aluminum	Aluminum

SOURCE: FactSet, "Financial Data and Analytics, Powered by Tomorrow."

Russia Sample Set

We generated a sample set for Russia from a variety of sources, including CNA, *The Drive*, the International Centre for Defense and Security, *Izvestia*, RIA Novosti, TASS, *Military Today*, and others. The Russian platforms chosen for the sample are analogous to the U.S. sample in terms of size and function.

Table B.8. Russian UxS Sample Set

User Type	Use Case	Type	Size	Future	Prime Manufacturer	Legacy
Commercial	Agricultural	UAV	Group 2	201 Agrogeo	GeoScan Group	
Commercial	Agricultural	UAV	Group 3	SKYF Agro	ARDN Technologies	
Commercial	Municipal	UAV	Group 1	Lite or Gemini	GeoScan Group	
Commercial	Search and rescue	UGS	Small (light)	Teledroid	Android Technologies	Fedor
Commercial	Maritime multiuse	USV	Small (Class 1)	Iskatel	Aviation and Marine Electronics	
Commercial	Sea floor mapping	USV	Large (Class 4)	CyberBoat-330	St. Petersburg Polytechnical University	
Commercial	Sea floor mapping	UUV	Medium	Marlin-350	Tetis Pro	
Commercial	Sea floor mapping	UUV	Large	Klavesin-2R-PM	Rubin Design Bureau (United Shipbuilding Corporation)	
Defense	Surveillance and reconnaissance	UAV	Group 3	Orlan-30	Special Technology Center	Orlan-10
Defense	Surveillance and reconnaissance	UAV	Group 5	Orion-2	Kronshtadt Group	Orion
Defense	Explosives disposal	UGS	Small (light)	Scorpion	SET-1	Scarab
Defense	Ground combat	UGS	Small (medium)	Soratnik & Nakhlebnik	Kalashnikov Concern	
Defense	Maritime	USV	Medium (Class 3)	Falco	Mortec	
Defense	Maritime	USV	Large (Class 4)	Poseidon	Rubin Design Bureau	
Defense	Maritime mine hunter	UUV	Medium	Vityaz-D	Rubin Design Bureau	
Defense	Maritime	UUV	Extra-large	Cephalopod	Rubin Design Bureau	
Recreational	Photo hobbyist	UAV	Group 2	401	GeoScan Group	
Recreational	Photo hobbyist	UAV	Group 1	S2	Optiplane	

International Sample Set

Finally, we developed an international set to represent the robust international community industry. We identified the top-selling defense systems using AUVSI data on system characteristics and Janes Markets Forecast data on international sales. For commercial systems, we identified key manufacturers and system characteristics using the AUVSI database, but in some cases we were not able to identify their most popular models.

Table B.9. International UxS Sample Set

User Type	Use Case	Type	Size	Model	Prime Manufacturer	Country
Commercial	Agricultural	UAS	Group 2	Disco-Pro AG	Parrot	France
Commercial	Agricultural	UAS	Group 3		Yamaha Motor Co.	Japan
Commercial	Municipal	UAS	Small (medium)		Draganfly SenseFly Uconsystem	Canada Switzerland Korea
Defense	Surveillance and reconnaissance	UAS	Group 3	Camcopter S-100	Schiebel	Austria
				Legacy: Patroller	Safran	France
Defense	Surveillance and reconnaissance	UAS	Group 5	Hermes 900	Elbit Systems	Israel
				Legacy: IAI Eitan (Heron TP)	Israel Aerospace Industries	Israel
Recreational	Photo hobbyist	UAS	Group 1	BeBop 2	Parrot	France
Recreational	Photo hobbyist	UAS	Group 2		Acecore Technologies	Netherlands
Commercial	Search and rescue	UGS	Small (light)	VersaTrax 320	Eddyfi Tech	Canada
Defense	EOD	UGS	Small (light)	MTGR	Roboteam	Israel
				Legacy: Gryf	PIAP	Poland
Defense	Ground combat	UGS	Small (medium)	THeMIS	Milrem	Estonia
				Legacy: tEODor	Unmanned Systems Investment GmbH (bought by AeroVironment)	Germany
Commercial	Maritime security	USV	Large (Class 4)	Yara Birkeland	Yara International	Norway
Commercial	Maritime security	USV	Small (Class 2)	DriX	iXblue	France
Defense	Maritime	USV	Large (Class 4)	Long-range autonomous naval vessel	Rolls-Royce	United Kingdom
Defense	Maritime	USV	Medium (Class 3)	Inspector	Group Gorge	France
				Legacy: Venus 16	Lung Teh Shipbuilding/ Singapore Technologies Electronics Limited	Taiwan/ Singapore
Commercial	Sea floor mapping	UUV	Large		Group Gorge/ECA	France
Commercial	Sea floor mapping	UUV	Medium	Legacy: AUV Fleet	Seamor Marine Subsea Tech Fugro	Canada France Netherlands
Defense	Maritime	UUV	Extra-large	Double Eagle	SAAB	Sweden
Defense	Maritime mine countermeasures	UUV	Medium	DTG2	Deep Trekker, Inc.	Canada

129

Appendix C. Interview Protocol

RAND is conducting a study that is exploring the state of the autonomous uncrewed systems industrial base for the Office of the Under Secretary of Defense for Research and Engineering (OUSD[R&E]). OUSD(R&E) wants to understand what risks, issues, and opportunities exist within the defense industrial base and the natural extensions to the broader uncrewed systems defense industrial base and the commercial and recreational systems base.

The types of systems include aircraft, ground, and maritime platforms, and the general time frame of focus is near term (i.e., five years), though relevant issues beyond that time frame are open for discussion.

We will not be attributing themes or quotes by name or position. We will list organizations that we spoke with in our final report. You are free to participate or not, and free to decline to answer any question.

Questions

1. Systems:
   a. What uncrewed system platforms are you currently procuring (i.e., buying new units this FY)?
   b. What uncrewed systems are you planning to procure in the near term?

2. For each system:
   a. Do you consider this system to be remote operated, automated, autonomous,[1] or some combination?
   b. If the system is remote operated, what are your plans for automation and autonomy?
   c. How many units have been procured historically?
   d. How many are you planning to procure in the near term?
      i. How certain are these plans, e.g., what is the risk of these planned quantities increasing or decreasing significantly?
   e. Where in the acquisition lifecycle is the platform, what is the TRL [technology readiness level]?
   f. Who is the prime integrator/contractor/builder?
   g. What are the subsystems for the platform?
      i. Who are the suppliers?
   h. Did you have competing vendors for the prime/competing suppliers who could produce the same components?
   i. What data rights and tech data packages does the U.S. government have for the system and its critical components?

---

[1] We generally define *autonomous* as choice making, whereas *automated* is preplanned.

3. Autonomy-enabling technologies:
   a. Which technologies do you consider essential to autonomous operation? Examples include sensors, navigation, guidance, command and control, communications, power, and energy.
   b. Who are the industry leaders for the identified technologies?

4. Industrial base criticality:

   Regardless of whether the system is autonomous or not, for each of the core elements of a UxS (hull and frame, propulsion and power, C3, autonomy enabling, sensors and payloads, and raw materials), we have the following questions:
   a. To what degree are alternative suppliers/capabilities available?
   b. To what degree is defense-specific knowledge needed to produce this capability?
   c. Is there a commercial market for this capability?
   d. Does producing this capability require specialized facilities or equipment?
   e. How long would it take to establish or expand this capability if it hasn't been developed yet (or past capacity is lost)?
   f. To what degree are specialized skills required to produce this capability?
   g. Are there raw materials, components, or steps in the manufacturing/integration process that have exceptionally long lead times?

5. Industrial base fragility:

   Regardless of whether the system is autonomous or not, for each of the core elements of the system (hull and frame, propulsion and power, C3, autonomy enabling, sensors and payloads, and raw materials), we have the following questions:
   a. Are suppliers dependent on DoD sales, or is there also a commercial market? (What fraction of the supplier's sales do DoD contracts account for?)
   b. What is the risk of the firm exiting the market?
   c. How many total firms are in the market?
   d. How reliant is this capability on non-U.S. components or materials?

6. Testing and evaluation:
   a. How many testing facilities support your system(s)?
   b. How many different tests are required for certification or qualification of the system(s)? How long does it take to get results from these tests?
   c. What is the capacity of these testing facilities? (i.e., What is the maximum number of units the test facility can support at one time? What is the typical throughput of these facilities?)
   d. Is there competing demand for these testing facilities? Have you experienced delays getting access to these facilities or getting results?

7. Risks and opportunities:
   a. What are the government (federal, state, local) regulatory or statutory policies in effect that are helping or hurting your acquisitions with respect to your engagement with industry and the health of the industrial base (e.g., environmental, test ranges, subsidies, public infrastructure requirements, product security requirements)?

b. Can you describe existing and potential policy changes to support your acquisitions?
   i. To support industry?
c. What trends do you see that will affect the uncrewed systems industry in the future?
d. Are there other risks, challenges, or opportunities related to the autonomous systems industrial bases that we have not identified here?

8. Conclusion:
   a. What questions are we not asking that we should be?
   b. Who else should we be speaking with?

Abbreviations

AGL	above ground level
AI	artificial intelligence
AUVSI	Association for Uncrewed Vehicle Systems International
AVIC	Aviation Industry Corporation
C2	command and control
C3	command, control, and communications
CAS	Chinese Academy of Sciences
CBRN	chemical, biological, radiological, or nuclear
CCL	Commerce Control List
COTS	commercial off-the-shelf
DARPA	Defense Advanced Research Projects Agency
DIB	defense industrial base
DIU	Defense Innovation Unit
DoD	Department of Defense
EOD	explosive ordnance disposal
EU	European Union
FAA	Federal Aviation Administration
FaC	fragility and criticality
FL	flight level
FY	fiscal year
GNSS	Global Navigation Satellite System
INS	inertial navigation system
ISR	intelligence, surveillance, and reconnaissance
LiDAR	light detection and ranging
LRRS	long-range reconnaissance and surveillance
MRM	medium-range mobile
NATO	North Atlantic Treaty Organization

NDAA	National Defense Authorization Act
ODASD(MIBP)	Office of the Deputy Assistant Secretary of Defense for Manufacturing and Industrial Base Policy
OSD	Office of the Under Secretary of Defense
OUSD(R&E)	Office of the Under Secretary of Defense for Research and Engineering
PLA	People's Liberation Army
R&D	research and development
RDT&E	research, development, test, and evaluation
SBS	soldier borne sensor
SIPRI	Stockholm International Peace Research Institute
SME	subject matter expert
SOCOM	Special Operations Command
SRM-UAS	short-range micro-UAS
STEM	science, technology, engineering, and mathematics
sUAS	small uncrewed aerial system
SURSS	Small Unit Remote Scouting System
TRL	technology readiness level
UAS	uncrewed aerial system
UGS	uncrewed ground system
USMC	U.S. Marine Corps
USML	United States Munitions List
USV	uncrewed surface vehicle
UUV	uncrewed underwater vehicle
UxS	uncrewed system
VTOL	vertical takeoff and landing
WBS	Work Breakdown Structure

References

Air Force Global Hawk Program Office, Air Force Materiel Command/Air Force Life Cycle Management Center, RQ-4 Global Hawk, interview with the authors, October 5, 2021.

Allied Market Research, "Unmanned Aerial Vehicle (UAV) Market by Type," webpage, October 2021. As of March 21, 2022:
https://www.alliedmarketresearch.com/unmanned-aerial-vehicle-market-A09059

Allik, Sten, Sean Fahey, Tomas Jermalavičius, Roger McDermott, and Konrad Muzyka, *The Rise of Russia's Military Robots: Theory, Practice and Implications*, Tallinn, Estonia: International Centre for Defence and Security, February 2021.

Alpine 4 Holdings, Inc., "Consolidated Statement of Operations," SEC Form 10-K for the year ended December 31, 2020, April 15, 2021. As of August 22, 2022:
https://www.sec.gov/cgi-bin/viewer?action=view&cik=1606698&accession_number =0001096906-21-000759&xbrl_type=v#

Arlt, Bodo, "Supply Chain Challenges of the Ferrite Industry as a Result of the Pandemic," *Bodo's Power Systems*, August 2021, pp. 14–15.

Army Robotic and Autonomous Systems Program Office, Program Executive Office Combat Support and Combat Service Support, interview with the authors, September 27, 2021.

Association for Uncrewed Vehicle Systems International, "Unmanned Vehicle Systems and Robotics Database: Air, Ground, and Sea," webpage, undated a. As of August 18, 2022:
http://robotdirectory.auvsi.org/home

———, "Who Is AUVSI?" webpage, undated b. As of February 23, 2022:
https://www.auvsi.org/who-auvsi

———, *2021 Defense Budget for Unmanned Systems and Robotics*, Arlington, Va., 2021a. As of November 14, 2021:
https://www.auvsi.org/dod-unmanned-systems-budget-report

———, "Funding for Unmanned Systems in the FY 2022 Defense Budget Request," *Assuring Autonomy Blog*, June 16, 2021b. As of November 14, 2021:
https://www.auvsi.org/industry-news/blog/funding-unmanned-systems-fy-2022-defense -budget-request

———, interview with the authors, September 16, 2021.

Austin, Lloyd J., III, "Remarks by Secretary of Defense Lloyd J. Austin III at the Reagan National Defense Forum (as Delivered)," transcript, U.S. Department of Defense, December 4, 2021. As of March 10, 2022:
https://www.defense.gov/News/Speeches/Speech/Article/2861931/remarks-by-secretary-of
-defense-lloyd-j-austin-iii-at-the-reagan-national-defen

AUVSI—*See* Association for Uncrewed Vehicle Systems International.

Avilés, Lirio, and Sally Sleeper, "Identifying and Mitigating the Impact of the Budget Control Act on High Risk Sectors and Tiers of the Defense Industrial Base: Assessment Approach to Industrial Base Risks," in *Proceedings of the Thirteenth Annual Acquisition Research Symposium:* Vol. I, *Wednesday Sessions*, Monterey, Calif.: Naval Postgraduate School, April 30, 2016, pp. 347–359.

Ball, Mike, "Additional Centaur UGVs to Provide EOD to US Military," Unmanned Systems Technology, May 18, 2021. As of March 10, 2022:
https://www.unmannedsystemstechnology.com/2021/05/flir-provides-additional-centaur
-ugvs-to-us-military

Beam Communications Holdings Limited, Annual Report for the Year Ending 20 June 2021, September 2021. As of August 22, 2022:
https://www.beamcommunications.com/investors/annual-reports

Benchmark Mineral Intelligence, "Benchmark's Lithium Carbonate Prices Reach New All-Time Highs," press release, London, October 6, 2021. As of March 10, 2022:
https://www.benchmarkminerals.com/membership/benchmarks-lithium-carbonate-prices
-reach-new-all-time-highs

Bisht, Inder Singh, "Russia Offers Orion-E Combat Drone for Export," *Defense Post*, June 29, 2021. As of March 10, 2022:
https://www.thedefensepost.com/2021/06/29/russia-orion-e-drone-export/

Boeing, interview with the authors, November 11, 2021.

Borak, Masha, and Yujie Xue, "How Shenzhen, the Hi-Tech Hub of China, Became the Drone Capital of the World," *South China Morning Post*, April 4, 2021. As of March 10, 2022:
https://www.scmp.com/tech/big-tech/article/3128004/how-shenzhen-hi-tech-hub-china
-became-drone-capital-world

Burton, Jason, "U.S. Geological Survey Releases 2022 List of Critical Minerals," press release, Reston Va.: U.S. Geological Survey, February 22, 2022. As of March 10, 2022:
https://www.usgs.gov/news/national-news-release/us-geological-survey-releases-2022-list
-critical-minerals

Валерий Бутымов [Valery Butymov], "новую морскую модульную платформу FALCO представили на форуме "Армия-2020" [The New Marine Modular Platform FALCO was Presented at the Forum "Army-2020"]," Mil. Press Flot Prom, August 29, 2020.

Chan, Minnie, "Drone Warfare Marks a First for SCO Drill, as Region Faces Up to Terror Attack Risks in Afghan Fallout," *South China Morning Post*, October 3, 2021. As of March 10, 2022:
https://www.scmp.com/news/china/military/article/3151041/drone-warfare-marks-first-sco-drill-region-faces-terror-attack

Chase, Michael S., Kristen A. Gunness, Lyle J. Morris, Samuel K. Berkowitz, and Benjamin S. Purser III, *Emerging Trends in China's Development of Unmanned Systems*, Santa Monica, Calif.: RAND Corporation, RR-990-OSD, 2015. As of March 10, 2022:
https://www.rand.org/pubs/research_reports/RR990.html

Chen, Stephen, "After Drones, China Turns to Unmanned Vessels to Boost Its Marine Power," *South China Morning Post*, December 5, 2013. As of March 10, 2022:
https://www.scmp.com/news/china/article/1373490/after-drones-china-turns-unmanned-vessels-boost-its-marine-power

Cheng Yu, "China Now 'Leader' in Unmanned Aircraft System," *China Daily*, September 1, 2020. As of March 9, 2022:
https://www.chinadailyhk.com/article/141869

Clover, Charles, "China Seeks Private Sector Help to Streamline Bloated Army," *Financial Times*, February 8, 2015.

Code of Federal Regulations, Title 14, Chapter 1, Subchapter F, Part 107, Small Unmanned Systems, 2021, last amended June 2, 2022. As of July 24, 2022:
https://www.ecfr.gov/current/title-14/chapter-I/subchapter-F/part-107

Code of Federal Regulations, Title 22, Chapter 1, Subchapter M, Part 121, The United States Munitions List, last amended June 30, 2022. As of July 24, 2022:
https://www.ecfr.gov/current/title-22/chapter-I/subchapter-M/part-121

Culpan, Tim, and Tae Kim, "Tech Sanctions Won't Sting Russia for a While," Bloomberg, February 25, 2022. As of March 2, 2022:
https://www.bloomberg.com/opinion/articles/2022-02-25/tech-sanctions-won-t-sting-russia-for-a-while

Defense Acquisition Improvement Act of 1990, 101st Congress, S.2916, 1990.

Defense Innovation Unit, interview with the authors, November 18, 2021.

Defense Science Board, *The Role of Autonomy in DoD Systems*, Washington, D.C.: U.S. Department of Defense, July 2012.

DIU—*See* Defense Innovation Unit.

Divis, Dee Ann, "Enthusiasm Builds for U.S. Military Drone Exports After Rule Change," *Inside Unmanned Systems*, July 11, 2019. As of August 27, 2021:
https://insideunmannedsystems.com/enthusiasm-builds-for-u-s-military-drone-exports-after-rule-change

DOD—*See* U.S. Department of Defense.

Doffman, Zak, "This New Coronavirus Spy Drone Will Make Sure You Stay Home," *Forbes*, March 5, 2020. As of March 10, 2022:
https://www.forbes.com/sites/zakdoffman/2020/03/05/meet-the-coronavirus-spy-drones-that-make-sure-you-stay-home

Draganfly, "Hybrid VTOL: Quantix™ Mapper," webpage, undated. As of March 10, 2022:
https://draganfly.com/products/quantix-mapper

———, "Draganfly Announces Record Revenue in Fourth Quarter and Fiscal 2020 Financial Results," press release, Vancouver, B.C., April 19, 2021. As of March 10, 2022:
https://www.globenewswire.com/news-release/2021/04/19/2212479/0/en/Draganfly-Announces-Record-Revenue-in-Fourth-Quarter-and-Fiscal-2020-Financial-Results.html

Duffie, Warren, Jr., "ONR Chief Unveils New Vision to Reimagine Naval Power," press release, Arlington Va.: Office of Naval Research, November 23, 2021. As of January 12, 2022:
https://www.navy.mil/Press-Office/News-Stories/Article/2851592/onr-chief-unveils-new-vision-to-reimagine-naval-power

Easton, Ian M., and L. C. Russell Hsiao, *The Chinese People's Liberation Army's UAV Project*, Arlington, Va.: Project 2049 Institute, OP 13-001, 2013.

Edmonds, Jeffrey, Samuel Bendett, Anya Fink, Mary Chesnut, Dmitry Gorenburg, Michael Kofman, Kasey Stricklin, and Julian Waller, *Artificial Intelligence and Autonomy in Russia*, Arlington, Va.: CNA, May 2021.

Einhorn, Bruce, "Combat Drones Made in China Are Coming to a Conflict Near You," *Bloomberg Businessweek*, March 18, 2021.

Episkopos, Mark, "The Russian Navy Loves Drones (And for Good Reason)," *National Interest*, June 22, 2021. As of March 2, 2022:
https://nationalinterest.org/blog/buzz/russian-navy-loves-drones-and-good-reason-188340

Experian Information Services, Inc., "Experian Commercial Credit Scores," data downloaded for selected companies through Nexis Uni, February 25 and March 8, 2022.

"Export Successes Are Vindicating Turkey's UAS Drive," *Aviation Week*, June 8, 2021. As of May 9, 2022:
https://aviationweek.com/defense-space/missile-defense-weapons/export-successes-are-vindicating-turkeys-uas-drive?msclkid=f5667f99cfea11ecaba6ced518eccd22

FAA—*See* Federal Aviation Administration.

FactSet, "Financial Data and Analytics, Powered by Tomorrow," homepage, undated. As of June 2, 2022:
https://www.factset.com/

Fal'tsman, V. K. "Import Substitution in the Economic Sectors of Russia," *Studies on Russian Economic Development*, Vol. 26, No. 5, 2015, pp. 452–459.

Federal Aviation Administration, *FAA Aerospace Forecast: Fiscal Years 2019–2039*, Washington, D.C., 2019. As of March 10, 2022:
https://www.faa.gov/data_research/aviation/aerospace_forecasts/media/FY2019-39_FAA_Aerospace_Forecast.pdf

Feist, Jonathan, "DJI Phantom 4 Pro Review—A Superb Airframe," DroneRush, April 21, 2021. As of July 25, 2022:
https://www.dronerush.com/dji-phantom-4-pro-review-11301/

———, "Best Commercial Drones—Surveying, Mapping, and Search & Rescue," DroneRush, March 23, 2022a. As of July 25, 2022:
https://dronerush.com/best-commercial-drones-14616

———, "Best Camera Drones: Photos from the Sky in Any Price Bracket," DroneRush, July 15, 2022b. As of July 25, 2022:
https://dronerush.com/best-camera-drone-4180

———, "Best Drones 2022: Affordable 5.4K Video Is a Reality!" DroneRush, July 18, 2022c. As of July 25, 2022:
https://dronerush.com/best-drones-1977

"First Export of Russian Orlan-E Drones Goes to Myanmar," DefenseWorld.net, January 22, 2021. As of March 10, 2022:
https://www.defenseworld.net/news/28818/First_Export_of_Russian_Orlan_E_Drones_Goes_to_Myanmar

Fisher, Jim, "The Best Drones for 2022," *PCMag*, last updated May 24, 2022. As of July 25, 2022:
https://www.pcmag.com/picks/the-best-drones

Forbes, "Sierra Nevada," webpage, undated. As of March 10, 2022:
https://www.forbes.com/companies/sierra-nevada/?sh=3f8eeef35d7b

Fortune Business Insights, *Unmanned Aerial Vehicle (UAV) Market Size, Share & COVID-19 Impact Analysis*, Maharashtra, India, July 2020. As of March 21, 2022: https://www.fortunebusinessinsights.com/industry-reports/unmanned-aerial-vehicle-uav -market-101603

French, Sally, "The Best Drones for Kids in 2022," The Drone Girl, January 7, 2022a. As of March 10, 2022: https://www.thedronegirl.com/2020/03/10/the-best-drones-for-kids-in-2020

———, "The Best Drones for Photographers of 2022," The Drone Girl, May 17, 2022b. As of July 26, 2022: https://www.thedronegirl.com/2020/03/23/best-drones-photographers-2020

Gartner, "Gartner Forecasts Global IoT Enterprise Drone Shipments to Grow 50% in 2020," press release, Egham, UK, December 4, 2019. As of February 24, 2022: https://www.gartner.com/en/newsroom/press-releases/2019-12-04-gartner-forecasts-global -iot-enterprise-drone-shipmen

General Dynamics Corporation, SEC Form 10-K for the fiscal year ended December 31, 2021, February 9, 2022. As of August 22, 2022: https://s22.q4cdn.com/891946778/files/doc_financials/2021/q4/2021-10-K-As-Filed -(2-9-2022).pdf

GeoScan Group, homepage, undated. As of March 10, 2022: https://www.geoscan.aero/en

———, "About Company," undated. As of August 30, 2022: https://www.geoscan.aero/en/about

Giegerich, Andy, and Suzanne Stevens, "Vigor Industrial Sold to East Coast Private Equity Firms," *Portland Business Journal*, July 25, 2019. As of March 10, 2022: https://www.bizjournals.com/portland/news/2019/07/25/vigor-industrial-sold-to-heady-east -coast-firm.html

Gilli, Andrea, and Mauro Gilli, "Why China Has Not Caught Up Yet: Military-Technological Superiority and the Limits of Imitation, Reverse Engineering, and Cyber Espionage," *International Security*, Vol. 43, No. 3, 2019, pp. 141–189.

Glass, Paulina, "Russia's Pistol-Packing Robot Is Scrambling for Parts," *Defense One*, February 28, 2019. As of July 25, 2022: https://www.defenseone.com/technology/2019/02/russias-pistol-packing-robot-scrambling -parts/155226/

Global Database, "Financial Statements of Cosworth Group Holdings Limited," 2022. As of August 24, 2022:
https://uk.globaldatabase.com/company/cosworth-group-holdings-limited

———, "Financial Statements of Valeport Limited," 2022. As of August 24, 2022:
https://uk.globaldatabase.com/company/valeport-limited

Goble, Paul, "Import Substitution in Russia Failing as Moscow Buys Products Not Technologies," *Eurasia Daily Monitor*, Vol. 16, No. 44, March 28, 2019. As of March 10, 2022:
https://jamestown.org/program/import-substitution-in-russia-failing-as-moscow-buys-products-not-technologies

Гумарова, Екатерина, [Gumarova, Ekaterina], "Если не станут мешать, через 10 лет российский рынок беспилотников будет не узнать [If They Don't Interfere, in 10 Years the Russian Drone Market Will Be Unrecognizable]," *Реальное Время* [*Real Time*], March 11, 2019.

Haensel, Brett, "Surfers, Swimmers, Boaters Run Into Summer-Disrupting Fiberglass Shortage," Bloomberg, August 30, 2021. As of July 25, 2022:
https://www.bloomberg.com/news/articles/2021-08-30/surfers-swimmers-boaters-run-into-summer-disrupting-shortage#xj4y7vzkg

Hambling, David, "Russia Enters Military Drone Export Market with Sale to Myanmar," *Forbes*, January 25, 2021a. As of March 10, 2022:
https://www.forbes.com/sites/davidhambling/2021/01/25/russia-enters-military-drone-export-market-with-sale-to-burma

———, "China's New Unmanned Attack Sub May Not Be What It Seems (Update: In Fact It's A Paper Tiger)," *Forbes*, July 9, 2021b. As of July 25, 2022:
https://www.forbes.com/sites/davidhambling/2021/07/09/chinas-new-unmanned-attack-sub-may-not-be-what-it-seems/?sh=37c0b89c3658

Hsu, Kimberly, *China's Military Unmanned Aerial Vehicle Industry*, Washington, D.C.: U.S.-China Economic and Security Review Commission, 2013. As of February 28, 2022:
https://www.uscc.gov/research/chinas-military-unmanned-aerial-vehicle-industry

Hull, Andrew W., David R. Markov, and Eric Griffin, *"Private" Chinese Aerospace Defense Companies*, Montgomery, Ala.: China Aerospace Studies Institute, Air University, 2021.

Huntington Ingalls Industries, 2020 Annual Report, February 11, 2021. As of August 22, 2022:
https://s29.q4cdn.com/772422961/files/doc_financials/2020/ar/2020-annual.pdf

Huntington Ingalls Industries/Hydroid, interview with the authors, November 16, 2021.

Industry Arc, "Unmanned Aircraft Systems (UAS) Market—Forecast (2022–2027)." As of March 21, 2022:
https://www.industryarc.com/Report/15014/unmanned-aircraft-systems-market.html

Innopolis University, "Innopolis University Develops and Autonomous Driving System for KAMAZ," press release, Innopolis, Russia, February 8, 2018. As of March 10, 2022:
https://media.innopolis.university/en/news/mic-rt-iu-kamaz/

International Trade Administration, "Unmanned Aircraft Systems," webpage, undated. As of January 24, 2022:
https://www.trade.gov/unmanned-aircraft-systems

Iridium Communications Inc., 2020 Annual Report, April 2021. As of August 22, 2022:
https://investor.iridium.com/annual-reports

Janes, "Unrivalled Trusted Intelligence," webpage, undated. As of February 23, 2022:
https://www.janes.com/about-janes/janes-defence-intelligence-tradecraft

———, *Janes Markets Forecast*, brochure, Washington, D.C., 2020. As of March 10, 2022:
https://cdn.ihs.com/www/pdf/Markets-Forecast-Brochure.pdf

———, "Janes Markets Forecast – UAV, UGV, USV Program Forecast." As of October 2021:
https://customer.janes.com/MarketsForecast/guided?view=chart&f=MARKET(Unmanned
%20Air%20Veh||Unmanned%20Gnd%20Veh||Unmanned%20Sea%20Veh||Unmanned
%20Systems)&pg=1

Jaskula, Brian W., *Lithium*, Washington, D.C.: U.S. Geological Survey, Mineral Commodity Summaries, January 2021. As of July 25, 2022:
https://pubs.usgs.gov/periodicals/mcs2021/mcs2021-lithium.pdf

Katz, Justin, "Navy Awards 'Overlord' Second-Phase Contract Mods to Undisclosed Companies," *Inside Defense*, October 2, 2019. As of March 10, 2022:
https://insidedefense.com/insider/navy-awards-overlord-second-phase-contract-mods
-undisclosed-companies

———, "For Unmanned Vessels, Navy Still Working Out Maintenance Strategy," *Breaking Defense*, January 12, 2022a. As of March 10, 2022:
https://breakingdefense.com/2022/01/for-unmanned-vessels-navy-still-working-out
-maintenance-strategy/

———, "SCO Ends Project Overlord, Shifts Unmanned Vessels to Navy," *Breaking Defense*, January 12, 2022b. As of March 10, 2022:
https://breakingdefense.com/2022/01/sco-ends-project-overlord-shifts-unmanned-vessels-to
-navy

L3 Harris, interview with the authors, October 22, 2021.

3

Leidos, interview with the authors, October 29, 2021.

Levada-Center, "Emigration," July 6, 2021. As of July 26, 2022:
https://www.levada.ru/en/2021/07/06/emigration/

Lin Long-xin and Zhang Bi-sheng, Technical Development and Operational Application of an Unmanned Surface Combat System, *Journal of Underwater Unmanned Systems* Vol. 26, No. 2, 2018, pp. 107–114.

Liu Caiyu, "Underwater Glider Haiyi 1000 Completes Mission," *Global Times*, October 15, 2017. As of July 26, 2022:
https://web.archive.org/web/20191127150316/https://www.globaltimes.cn/content/1070405.shtml

Liu Haijiang, Li Xiangang, and Liang Ming, "A Thinking on Accelerating the Development of Unmanned Weapon Systems and Technologies," *National Defense Technology*, Vol. 41, No. 6, December 2020, pp. 32–36.

Liu Xuanzun, "China's First Unmanned Missile Boat Revealed at Airshow China 2018," *Global Times*, November 7, 2018. As of July 26, 2022:
https://web.archive.org/web/20200307005606/http://www.globaltimes.cn/content/1126362.shtml

———, "PLA Commissions New-Type Robot for Bomb Disposal," *Global Times*, March 31, 2021a. As of March 9, 2022:
https://www.globaltimes.cn/page/202103/1219983.shtml

———, "Self-Made Recon Robot Tested for Coast Patrol, Shows PLA's 'Grassroots Innovation for Modernization,'" *Global Times*, June 1, 2021b. As of March 9, 2022:
https://www.globaltimes.cn/page/202106/1225106.shtml

Maizland, Lindsay, and Andrew Chatzky, *Huawei: China's Controversial Tech Giant*, New York: Council on Foreign Relations, August 6, 2020. As of February 28, 2022:
https://www.cfr.org/backgrounder/huawei-chinas-controversial-tech-giant

Markets and Markets, "Unmanned Aerial Vehicle (UAV) Market by Point of Sale, Systems, Platform (Civil & Commercial, and Defense & Government), Function, End Use, Application, Type, Mode of Operation, MTOW, Range, and Region (2021–2026)," June 2021. As of March 21, 2022:
https://www.marketsandmarkets.com/Market-Reports/unmanned-aerial-vehicles-uav-market-662.html

Marson, James, and Giovanni Legorano, "China Bought Italian Military-Drone Maker Without Authorities' Knowledge," *Wall Street Journal*, November 15, 2021.

Mason, Hannah, "Resin Shortages Continue to Affect Composites Supply Chain," *CompositesWorld*, May 3, 2021. As of February 25, 2022:
https://www.compositesworld.com/news/resin-shortages-continue-to-affect-composites-supply-chain

Meaker, Morgan, "Russia's War in Ukraine Could Spur Another Global Chip Shortage," *Wired*, February 28, 2022. As of March 14, 2022:
https://www.wired.com/story/ukraine-chip-shortage-neon

MIND Technology, Inc., SEC Form 10-K for the fiscal year ended January 31, 2021.

Mönch Publishing Group, "Russia to Receive New-Generation UAS by 2030," February 25, 2020. As of March 10, 2022:
https://monch.com/russia-to-receive-new-generation-uas-by-2023/

Monterey Bay Aquarium Research Institute, "Seafloor Mapping AUV," undated. As of March 10, 2022:
https://www.mbari.org/at-sea/vehicles/autonomous-underwater-vehicles/seafloor-mapping-auv

Nature Index, "2021 Tables: Institutions," webpage, undated. As of February 28, 2022:
https://www.natureindex.com/annual-tables/2021/institution/all/all/global

Naval Drones, "Tianxiang One USV," webpage, undated. As of March 10, 2022:
http://www.navaldrones.com/Tianxiang-One.html

Navy and Marine Corps Small Tactical Unmanned Aircraft Systems Program Office, Naval Air Systems Command/PMA-263, interview with the authors, September 30, 2021.

Navy Littoral Combat Ship Mission Modules Program Office, Program Executive Office Unmanned and Small Combatants, Naval Sea Systems Command/PMS-420, interview with the authors, November 23, 2021.

Navy Unmanned Maritime Systems Program Office, Program Executive Office Unmanned and Small Combatants, Naval Sea Systems Command/PMS-406, interview with the authors, October 1, 2021.

Newdick, Thomas, and Tyler Rogoway, "Russia's Predator-Style Drone with Big Export Potential Has Launched Its First Missiles," *The Drive*, December 28, 2020. As of March 10, 2022:
https://www.thedrive.com/the-war-zone/38446/russias-predator-style-drone-with-big-export-potential-has-launched-its-first-missiles

Nurkin, Tate, Kelly Bedard, James Clad, Cameron Scott, and Jon Grevatt, *China's Advanced Weapons Systems*, London: Jane's by IHS Markit, May 12, 2018.

Office of the Deputy Assistant Secretary of Defense for Systems Engineering, *Department of Defense Risk, Issue, and Opportunity Management Guide for Defense Acquisition Programs*, Washington, D.C., January 2017.

Office of Management and Budget, *A Budget for America's Future: Budget of the U.S. Government, Fiscal Year 2021*, Washington, D.C.: U.S. Government Publishing Office, 2020. As of November 14, 2021:
https://www.govinfo.gov/content/pkg/BUDGET-2021-BUD/pdf/BUDGET-2021-BUD.pdf

Office of the Secretary of Defense, *Military and Security Developments Involving the People's Republic of China*, Washington, D.C., 2020.

Ohlandt, Chad J. R., and Jon Schmid, "Flying High: Chinese Innovation in Unmanned Aerial Vehicles," paper presented at the CAPS-RAND-NDU Conference on the People's Liberation Army, Arlington, Va., April 30, 2018.

Операция «Преемник» идет полным ходом—робота Федора на МКС заменит «Теледроид» [Operation "Successor" Is in Full Swing—Fedor's Robot on the ISS Will Be Replaced by "Teledroid"], *Meduza*, February 8, 2022.

O'Rourke, Ronald, *Navy Large Unmanned Surface and Undersea Vehicles: Background and Issues for Congress*, Washington, D.C.: Congressional Research Service, R45757, March 17, 2021.

OSD—*See* Office of the Secretary of Defense.

Oshkosh Corporation, SEC Form 10-K for the fiscal year ended September 30, 2021, November 16, 2021. As of August 22, 2022:
https://s24.q4cdn.com/975203404/files/doc_financials/2021/ar/Oshkosh_AR21_update_35243.pdf

Peck, Michael, "Coming Soon: World-Class Chinese Military Drones," *National Interest*, July 10, 2021a. As of January 24, 2022:
https://nationalinterest.org/blog/reboot/coming-soon-world-class-chinese-military-drones-189498

———, "This Massive Russian Sub Is Preparing to Launch Nuclear Torpedoes," *National Interest*, September 17, 2021b. As of July 26, 2022:
https://nationalinterest.org/blog/reboot/massive-russian-sub-preparing-launch-nuclear-torpedoes-193673

Philip, Snehesh Alex, "China Has Become a Major Exporter of Armed Drones, Pakistan Is Among Its 11 Customers," ThePrint, November 23, 2020. As of August 27, 2021:
https://theprint.in/defence/china-has-become-a-major-exporter-of-armed-drones-pakistan-is-among-its-11-customers/549841/

Phoenix International Holdings, Inc., "Phoenix International Holdings, Inc. (Phoenix) Has Been Contracted by the Commonwealth of Australia Under the Direction of Australian Transport Safety Bureau (ATSB) to Provide Deepwater Towed Side Scan Sonar Services," press release, Largo, Md., January 29, 2016. As of March 10, 2022: https://www.phnx-international.com/phoenix-international-returns-to-search-for-malaysia-airlines-flight-370-mh370

Pilot Institute, "Airworthiness Certification of Drones—Everything You Need to Know," July 18, 2021. As of February 25, 2022: https://pilotinstitute.com/drone-airworthiness-certification

Precedence Research, "UAV Drones Market Size to Worth Around US$ 102.38 Bn by 2030," March 1, 2022. As of March 21, 2022: https://www.globenewswire.com/news-release/2022/03/01/2394915/0/en/UAV-Drones-Market-Size-to-Worth-Around-US-102-38-Bn-by-2030.html

Вячеслав Прокофьев [Vyacheslav Prokofyev], "РБК: "Сколково" и "Роснано" перейдут под управление ВЭБ.РФ [RBC: Skolkovo and Rosnano will come under the control of VEB.RF]," Tass, November 22, 2020.

Rapa, Anthony, "From Unmanned Systems Magazine: Unmanned Systems and Export Controls: What Your Company Needs to Know," Association for Unmanned Vehicle Systems International, May 21, 2018. As of January 24, 2022: https://www.auvsi.org/unmanned-systems-magazine-unmanned-systems-and-export-controls-what-your-company-needs-know

Ray, Jonathan, Katie Atha, Edward Francis, Caleb Dependahl, James Mulvenon, Daniel Alderman, and Leigh Ann Ragland-Luce, *China's Industrial and Military Robotics Development*, Washington, D.C.: U.S.-China Economic and Security Review Commission, October 2016.

Raytheon Technologies Corporation, SEC Form 10-K for the fiscal year ended December 31, 2021, February 8, 2021. As of August 24, 2022: https://sec.report/Document/0000101829-21-000008/

Reid, David, "A Swarm of Armed Drones Attacked a Russian Military Base in Syria," CNBC, January 11, 2018. As of March 10, 2022: https://www.cnbc.com/2018/01/11/swarm-of-armed-diy-drones-attacks-russian-military-base-in-syria.html

Robotic Systems Joint Project Office, *Unmanned Ground Vehicle (UGV) Interoperability Profile (IOP) Overarching Profile*, Warren, Mich., December 21, 2011.

Rolls-Royce Holdings plc, Annual Report 2020, 2021. As of August 24, 2022: https://www.rolls-royce.com/~/media/Files/R/Rolls-Royce/documents/annual-report/2020/2020-full-annual-report.pdf

Rostec, Annual Report of the Rostec State Corporation 2019, Science, Overcoming Technologic Barriers, approved by the Supervisory Board of the Rostec State Corporation April 30, 2020.

Rozoff, Rick, "Turkey's Arms Sales to Ukraine up 30 Times in First Quarter over Last Year," *Daily Sabah*, April 6, 2022. As of May 9, 2022: https://antibellum679354512.wordpress.com/2022/04/06/turkeys-arms-sales-to-ukraine-up -30-times-in-first-quarter-over-last-year/?msclkid=f56545aecfea11ecb4c7669bef5f7fa3

Russian Advanced Research Fund, press releases.

"Russian Drone Orlan-10 Consists of Parts Produced in the USA and Other Countries—Photo Evidence," Inform Napalm, February 6, 2018. As of March 10, 2022: https://informnapalm.org/en/russian-drone-orlan-10-consists-of-parts-produced-in-the-usa -and-other-countries-photo-evidence

"Russia's State Arms Exporter to Offer Kamikaze Drones, Heavy UAVs to Foreign Customers," TASS, November 25, 2021. As of July 26, 2022: https://tass.com/defense/1366093

SberBank, "Cognitive Pilot," webpage, undated. As of March 10, 2022: https://www.sberbank.com/eco/cognitivepilot

SET-1, "About Us," webpage, undated. As of March 10, 2022: https://en.set-1.ru/about/

Shan, Xiuxu, *Research on Trade in Military Products of ZYD Institute*, Zhengzhou, China: Zhengzhou University, 2016.

Silberglitt, Richard, James T. Bartis, Brian G. Chow, David L. An, and Kyle Brady, *Critical Materials: Present Danger to U.S. Manufacturing*, Santa Monica, Calif.: RAND Corporation, RR-133-NIC, 2013. As of February 24, 2022: https://www.rand.org/pubs/research_reports/RR133.html

SIPRI—*See* Stockholm International Peace Research Institute.

Skydio, "Skydio 2," webpage, undated. As of December 15, 2021: https://www.skydio.com/skydio-2

———, interview with the authors, October 26, 2021.

Sleeper, Sally, and John F. Starns, *Implementing Filters to Identify and Prioritize Industrial Base Risk: Rules of Thumb to Reduce Cognitive Overload*, Washington, D.C.: Office of the Deputy Assistant Secretary of Defense for Manufacturing and Industrial Base Policy, ADA623501, 2015.

Sleeper, Sally, Gene Warner, and John Starns, "Identifying and Mitigating Industrial Base Risk for the DoD: Results of a Pilot Study," in *Proceedings of the Eleventh Annual Acquisition Research Symposium*: Vol. II, *Thursday Sessions*, Monterey, Calif.: Naval Postgraduate School, April 30, 2014, pp. 156–171.

Statista, "Enterprise Drone Unit Shipments Worldwide from 2020 to 2030 (in Millions)," February 14, 2022. As of February 24, 2022:
https://www.statista.com/statistics/1234569/worldwide-enterprise-drone-market-shipments

Strachan, David R., "China Enters the UUV Fray," *The Diplomat*, November 22, 2019. As of March 10, 2022:
https://thediplomat.com/2019/11/china-enters-the-uuv-fray/

Stockholm International Peace Research Institute, "Top List TIV Tables," web database, undated. As of March 10, 2022:
https://armstrade.sipri.org/armstrade/page/toplist.php

Sun Yongsheng, Jin Wei, and Tang Yuchao, "Application of Unmanned System in Prevention and Control of COVID-19," *Science and Technology Review* (China), Vol. 38, No. 4, 2020, pp. 39–49.

Sutton, H. I., "Chinese Navy Crafts Unmanned Sea Hunter Knock-Off," *USNI News*, September 25, 2020. As of March 10, 2022:
https://news.usni.org/2020/09/25/chinese-navy-crafts-unmanned-sea-hunter-knock-off

Teledyne Technologies, 2020 Annual Report, February 25, 2021. As of August 22, 2022:
https://www.teledyne.com/en-us/investors/Documents/Teledyne%20Technologies%20Incorporated%202020%20Annual%20Report.pdf

Trefis Team, "Are Battery Cost Improvements Still a Big Driver of Tesla's Margins?" *Forbes*, December 1, 2021. As of February 25, 2022:
https://www.forbes.com/sites/greatspeculations/2021/12/01/are-battery-cost-improvements-still-a-big-driver-of-teslas-margins

Trevithick, Joseph, "This U.S. Army Manual Has New Official Details About the RQ-170 Sentinel Drone," *The Drive*, May 12, 2017. As of March 10, 2022:
https://www.thedrive.com/the-war-zone/10221/this-u-s-army-manual-has-new-official-details-about-the-rq-170-sentinel-drone

Triumph Group, 2021 Sustainability and Annual Report, 2021. As of August 24, 2022:
https://s23.q4cdn.com/323685665/files/doc_financials/2021/ar/TGI-1417_21-Sustainability-Annual-Report-FINAL-Spreads.pdf

UAS Task Force, Airspace Integration Integrated Product Team, *Unmanned Aircraft System Airspace Integration Plan*, version 2.0, Washington, D.C.: U.S. Department of Defense, March 2011.

U.S. Department of Commerce, Bureau of Industry and Security, Export Administration Regulations, Supplement No. 1 to Part 774, Category 8—Marine, September 11, 2020. As of January 24, 2022:
https://www.bis.doc.gov/index.php/documents/regulations-docs/federal-register-notices/federal-register-2014/954-ccl8/file

———, Export Administration Regulations, Supplement No. 1 to Part 774, Category 9—Aerospace and Propulsion, March 29, 2021. As of January 24, 2022:
https://www.bis.doc.gov/index.php/documents/regulations-docs/2340-ccl9-4/file

———, Export Administration Regulations, Supplement No. 1 to Part 774, Category 0—Nuclear Materials, Facilities, and Equipment (and Miscellaneous Items), February 3, 2022. As of July 26, 2022:
https://www.bis.doc.gov/index.php/documents/regulations-docs/2331-category-0-nuclear-materials-facilities-equipment-and-miscellaneous-items-1/file

U.S. Department of Defense, *Unmanned Systems Integrated Roadmap, 2013–2038*, Washington, D.C., August 28, 2014.

———, *Unmanned Systems Integrated Roadmap, 2017–2042*, Washington, D.C., August 28, 2018.

———, *Counter–Small Unmanned Aircraft Systems Strategy*, Washington, D.C., 2020a.

———, *Department of Defense Standard Practice Work Breakdown Structures for Defense Materiel Items*, Washington, D.C., MIL-STD-881E, October 6, 2020b.

———, "Ghost Fleet Overlord Unmanned Surface Vessel Program Completes Second Autonomous Transit to the Pacific," press release, Washington, D.C., June 7, 2021. As of March 10, 2022:
https://www.defense.gov/News/Releases/Release/Article/2647818/ghost-fleet-overlord-unmanned-surface-vessel-program-completes-second-autonomou/

U.S. Department of State, "U.S. Policy on the Export of Unmanned Aerial Systems," fact sheet, May 21, 2019. As of July 2, 2021:
https://www.state.gov/u-s-policy-on-the-export-of-unmanned-aerial-systems

U.S. Government Accountability Office, *Nonproliferation: Agencies Could Improve Information Sharing and End-Use Monitoring on Unmanned Aerial Vehicle Exports*, Washington, D.C., GAO-12-536, July 30, 2012.

Vayu Aerospace, interview with the authors, September 20, 2021.

Venture Collective, "The Valley of Death," undated. As of May 13, 2022:
https://www.theventurecollective.com/the-valley-of-death/

Voronov, Vladimir, *Import Substitution for Rogozin*, trans. Arch Tait, London: Henry Jackson Society, January 2016. As of March 10, 2022:
http://henryjacksonsociety.org/wp-content/uploads/2018/06/1601-Import-Substitution-for-Rogozin.pdf

Wang Wen-feng, Yu Xue-mei, and Xu Dong-mei, "Overview of Unmanned Systems Interoperability Standardization," *China Electronic Standardization Institute*, Vol. 12, No. 1, 2020, pp. 100–104.

Wang Xingcheng and Chen Hai, "Application of Unmanned Combat System and Research on Key Issues," *Military Digest*, Vol. 4, April 2021, pp. 22–27.

White House, "Fact Sheet: Joined by Allies and Partners, the United States Imposes Devastating Costs on Russia," February 24, 2022. As of July 26, 2022:
https://www.whitehouse.gov/briefing-room/statements-releases/2022/02/24/fact-sheet-joined-by-allies-and-partners-the-united-states-imposes-devastating-costs-on-russia/

Wilson, J. R., "Unmanned Submarines Seen as Key to Dominating the World's Oceans," *Military and Aerospace Electronics*, October 15, 2019. As of March 10, 2022:
https://www.militaryaerospace.com/unmanned/article/14068665/unmanned-underwater-vehicles-uuv-artificial-intelligence

World Bank, "Research and Development Expenditure (% of GDP)—Russian Federation," graph, data as of June 22. As of July 26, 2022:
https://data.worldbank.org/indicator/GB.XPD.RSDV.GD.ZS?end=2020&locations=RU&start=1996

XAG, "XAG XPlanet Agricultural UAS: The Ultimate Agricultural Drone," webpage, undated. As of March 10, 2022:
https://www.xa.com/en/xp2020

Yang, Shulin, Xiaobing Yang, and Jianyou Mo, "The Application of Unmanned Aircraft Systems to Plant Protection in China," *Precision Agriculture*, Vol. 19, No. 2, April 2018, pp. 278–292.

Yang Wei, Wang Yue, Liu Xuechao, Zhao Kai, Xue Peng, and Zhang Bo, "Development and Research Summary of Unmanned Air Defense Weapons," *Journal of Gun Launch and Control*, March 6, 2021, pp. 1–7.

Заквасин, Алексей [Zakvasin, Aleksei], and Елизавета Комарова [Elizaveta Komarova], "«Технологическая состоятельность»: руководители «Кронштадта» — о возможностях БПЛА «Орион» и планах по развитию компании" ["Technological Viability": The Leaders of "Kronstadt"—About the Capabilities of the UAV 'Orion' and Plans for the Development of the Company], *RT*, October 10, 2021.

Zhang Xishan, Lian Guangyao, Li Huijie, and Yuan Xiangbo, "Development and Application of Intelligent Unmanned Support Equipment," *National Defense Technology*, Vol. 41, No. 2, April 2020, pp. 10–14.

Zhen, Liu, and Kristin Huang, "China's High-Speed, Armed Reconnaissance Drone Completes Maiden Flight," *South China Morning Post*, January 13, 2021. As of March 9, 2022: https://www.scmp.com/news/china/military/article/3117569/chinas-high-speed-armed -reconnaissance-drone-completes-maiden

Zumbrun, Josh, and Alex Leary, "Chip Shortage Leaves U.S. Companies Dangerously Low on Semiconductors, Report Says," *Wall Street Journal*, January 25, 2022.